2007 6th IEEE Dallas Circuits and Systems Workshop on System-on-Chip

Dallas, TX
15-16 November 2007

IEEE Catalog Number: CFP07505-PRT
ISBN 10: 1-4244-1679-5
ISBN 13: 978-1-4244-1679-0

**Copyright © 2007 by The Institute of Electrical and Electronics Engineers, Inc.
All Rights Reserved**

Copyright and Reprint Permissions: Abstracting is permitted with credit to the source. Libraries are permitted to photocopy beyond the limit of U.S. copyright law for private use of patrons those articles in this volume that carry a code at the bottom of the first page, provided the per-copy fee indicated in the code is paid through Copyright Clearance Center, 222 Rosewood Drive, Danvers, MA 01923.

For other copying, reprint or republications permission, write to IEEE Copyrights Manager, IEEE Operations Center, 445 Hoes Lane, Piscataway, New Jersey USA 08854. All rights reserved.

IEEE Catalog Number:	CFP07505-PRT
ISBN 10:	1-4244-1679-5
ISBN 13:	978-1-4244-1679-0
LOC:	2007906917

Additional Copies of This Publication Are Available from:

IEEE Service Center
445 Hoes Lane
Piscataway, NJ 08854

Phone:	(800) 678-IEEE
	(732) 981-1393
Fax:	(732) 981-9667
E-mail:	customer-service@ieee.org

2007 6th IEEE Dallas Circuits and Systems Workshop on System-on-Chip

Dallas, Texas
15-16 November 2007

IEEE Catalog Number: CFP07505-POD
ISBN: 978-1-42441-679-0

Foreword

System-on-Chips that integrates large-scale digital, analog, and mixed-signal components and cores continue to drive today's semiconductor industry. The Dallas Chapter of the IEEE Circuits and Systems Society organized the Sixth Dallas Circuits and Systems (DCAS-07) Workshop to address issues related to the design of such circuits and systems. This volume contains the collection of contributions presented at this workshop on System-on-Chip: Design, Application, Integration, and Software.

The purpose of this workshop is to provide an excellent opportunity for participants to exchange information and keep abreast with current trends in circuits and systems through presentations and discussions with leading experts in the field from industry and academia.

After the success of the previous five DCAS workshops, this year the organizing committee decided to continue the format of a two-day workshop from last year. This enabled us to include paper presentations besides the popular invited presentations and poster sessions. This year we have five distinguished speakers who will present leading-edge topics in advanced electronics. Additionally, we will have eight paper presentations from the papers submitted to the workshop. We also continue the successful poster sessions first introduced in DCAS-01.

We are grateful to our distinguished invitees, Prof. Michael Perrott (Massachusetts Institute of Technology), Prof. Bora Nikolic (University of California, Berkeley), Dr. Ken Kundert (Designers-Guide Consulting), and Dr. Kush Gulati (Cambridge Analog Technologies), and to our keynote speaker Dr. Dennis Buss (Texas Instruments Inc.). I would also like to thank the organizing and technical committees for working so hard to make the 6th IEEE DCAS workshop successful.

Dallas, November 2007

Ping Gui
General Chair

Organizing Committee

General Chair
Peggy Ping Gui
Southern Methodist University
Dallas TX, 75257
Email: pgui@engr.smu.edu

General Co-Chair
Dr. Mitch Thornton
Southern Methodist University
Dallas TX, 75257
Email: mitch@engr.smu.edu

Local Arrangements Chair
Arjun Rajagopal
Texas Instruments, Inc.
MS 8635
P.O BOX 650311
Dallas TX, 75265
Email: arjun@ti.com

Treasurer
Mak Kulkarni
Texas Instruments Inc.
MS 366,
P.O. Box 655012
Dallas TX, 75265
Email: mak@ti.com

IEEE DCAS Chair
Liming Xiu
Texas Instruments Inc.
MS 8710
P.O BOX 650311
Dallas TX, 75265
Email: limingxiu@ti.com

Technical Program Chair
Robert Bogdan Staszewski
Texas Instruments, Inc.
MS 8728
P.O BOX 650311
Dallas TX, 75265
Email: b-staszewski@ti.com

Publications Chair
Sudhind Dhamankar
Texas Instruments, Inc.
MS 8728
P.O BOX 650311
Dallas TX, 75265
Email: sudhind@ti.com

Publicity Chair
Dr Dinesh Bhatia
University of Texas at Dallas
P.O BOX 830688
Richardson TX, 75083
Email: dinesh@utdallas.edu

Registration Chair
Luke Wu
Texas Instruments Inc.
MS 8710
P.O BOX 650311
Dallas TX, 75265
Email: limingxiu@ti.com

System-on-Chip (SoC): Design, Application, Integration and Software

Second Call for Papers

Sixth IEEE Dallas Circuits and Systems Workshop

Southern Methodist University
Dallas, TX, USA

Nov 15-16, 2007

The IEEE Circuits and System Society Dallas Section is conducting its sixth workshop (DCAS-07) to provide a forum for sharing design, application, integration and software aspects of System-on-Chip (SoC).

DCAS-07 is sponsored by IEEE Circuits and System Society Dallas Section and will be held on Nov. 15-16 on the campus of Southern Methodist University, Dallas TX

The technical program committee invites researchers from academia and industry to contribute new and previously unpublished results in the following Circuits and Systems related areas:

- High Performance and Low Power Circuits
- Digital Signal Processors and Cores
- Baseband Communication Processors
- RF Processors and Circuits
- Multimedia Processors
- Reconfigurable Processors
- Low Voltage and Mixed Signal Circuits
- Mixed Signal Integration

- A/D and D/A Conversion
- High Speed I/O
- Clocking and Clock Distribution
- Power Management Circuits
- Signal Integrity and On-chip Interconnects
- SoC Implementation Methodology
- SoC Verification
- SoC and IP integration

Paper Submission Details

Submission deadline: now extended to 08/31/2007
Acceptance notice: 09/28/2007
Camera ready draft due: 10/12/2007

Web-based, automated paper submissions (not to exceed 4 pages), review and notification process
http://www.ewh.ieee.org/soc/cas/dallas/wks2007

Organizing Committee

General Chair: Ping Gui, SMU
General Co-Chair: Mitch Thornton, SMU
Technical Program Chair: Bogdan Staszewski, Texas Instruments
Publicity Chair: Dinesh Bhatia, UT Dallas
Publications Chair: Sudhind Dhamankar, Texas Instruments
Local Arrangements Chair: Arjun Rajagopal, Texas Instruments
Registration Chair: Luke Wu, Texas Instruments
Treasurer: Mak Kulkarni, Texas Instruments
IEEE-DCAS Chair: Liming Xiu, Texas Instruments

Technical Program Committee

Bogdan Staszewski, Technical Program Chair, TI
Mitch Thornton, Technical Program Co-Chair, SMU
Poras T. Balsara, Technical Program Co-Chair, UT Dallas
Oren Eliezer, TI
Terry Blake, TI
Ranjit Gharpurey, UT Austin
Sebastian Hoyos, Texas A&M
Mak Kulkarni, TI
Donald Lie, Texas Tech Univ.
Saraju P. Mohanty, UNT
Vojin Oklobdzija, UT Dallas
Weiping Shi, Texas A&M
Naveen Yanduru, TI

Ping Gui, SMU
Gernot Hueber, DICE, Austria
Hoi Lee, UT Dallas
Andrew Marshall, TI
Nagi Nagamathan, LSI
Kaijian Shi, Synopsys
Liming Xiu, TI

http://www.ewh.ieee.org/soc/cas/dallas/wks2007

Program for the 6th IEEE Dallas/CAS Workshop (DCAS-07)

System-on-Chip (SoC): Applications, Integration, Implementation and Software

THURSDAY NOVEMBER 15, 2007

7:30-8:30 AM	*Poster Setup*
8:30 AM	Opening Remarks
8:40 AM	*Keynote Address :*
	Si Technology Roadmap For Ubiquitous Computing, Sensing, and Perception
	Dr. Dennis Buss
	Texas Instruments, Dallas, TX, U.S.A.
10:10 AM	*BREAK*
10:30 AM	*Invited Talk 1 :* **Making Better Use of Time in Mixed-Signal Circuits**
	Prof. Michael Perrott
	Massachusetts Institute of Technology (MIT), U.S.A.
12:00 PM	*LUNCH*
1:30 PM	*Invited Talk 2 :* **Reconfigurable ADC**
	Dr. Kush Gulati
	Bitwave Semiconductor, U.S.A.
3:00PM	**Linearization of Highly-Efficient Monolithic Class E SiGe Power Amplifiers with Envelope-Tracking (ET)**
	Donald Y.C. Lie, J.D. Popp, F. Wang*, D. Kimball*, and L.E. Larson**
	Texas Tech University, *University of California, San Diego (UCSD), U.S.A.
3:20 PM	**Digital FIR Filter Optimization Using Toggle-Based Power Estimation Tools**
	Albina Cristian and Gunther Hackl
	GME mbH, Germany
3:40 PM	*BREAK*

THURSDAY POSTERS

3:40 PM	**Layout Parasitic Interconnections Effects on High Frequency Circuits**
	Albina Cristian and Gunther Hackl
	GME mbH, Germany
	"Flying-Adder" PLL Based Synchronization Mechanism for Data Packet Transport
	Liming Xiu, Steve Clynes, Srikanth Gurrapu, Towfique Haider, Feng Ying, and Wahed Mohammed
	Texas Instruments Inc., U.S.A.
	System-On-Chip Power Consumption Refinement and Analysis
	David Feinstein, Mitch Thornton, and Faith Kocan
	Southern Methodist University, U.S.A.
	A High-Performance Multi-Match Priority Encoder for TCAM-Based Packet Classifiers
	Miad Faezipour, Mehrdad Nourani
	University of Texas at Dallas, U.S.A.
	An Efficient Implementation of Scalable Architecture for Discrete Wavelet Transform On FPGA
	Rabah Hassan, Michael Guarisco, Xun Zhang, and Serge Weber
	Nancy University, France
	A New Fast Slew Buffering Algorithm Without Input Slew Assumptions
	Shiyan Hu and Jiang Hu
	Texas A&M University, U.S.A.
	Closed form Equations for Inter-modulation Distortion Parameters in WCDMA Receiver Validated Through Measurements
	*Mohammed Saif Khan and Naveen K. Yanduru**
	Texas Instruments, Bangalore, India. *Texas Instruments, Dallas, TX, USA.
	A CMOS Wideband LNA Using Multiple Phase Matched Frequency Staggered Resonators
	Diptendu Ghosh and Ranjit Gharpurey
	University of Texas, Austin, U.S.A.
5:00 PM	*Poster Removal*

vi

FRIDAY, NOVEMBER 16, 2007

7:30-8:50 AM	*Poster Setup*

8:50AM	*Invited Talk 3 : Verification of Complex Analog Integrated Circuits*

Dr. Ken Kundert
Designers-Guide Consulting, U.S.A.

10:20AM	*BREAK*

10:40AM	**Analysis of Third-order Intermodulation in Receiver Down-converter with Multiband Feedback**

Junghwan Han and Ranjit Gharpurey
University of Texas, Austin, U.S.A.

11:00 AM	**Enhancement of Coexistence Performance in a DRP Based Multi-Radio Environment of a Mobile Phone**

Yossi Tsfati, Oren Eliezer, Ran Katz, Yaniv Tzoret, and Ofer Friedman
Texas Instruments Inc., Wireless Terminals Business Unit, Raanana, Israel.

11:20 AM	**Waveform Analysis and Delay Prediction in Simultaneously Switching CMOS Gate Driven Inductively and Capacitively Coupled On-Chip Interconnects**

B.K. Kaushik, S. Sarkar, R.P. Agarwal, and R.C. Joshi
Indian Institute of Technology-Roorkee, *MITS-Sikar, INDIA

11:40 AM	**Design of a Multi-Context FPVLSI based on an Asynchronous Bit-Serial Architecture**

Waidyasooriya Hasitha Muthumala, Masanori Hariyama, and Michitaka Kameyama
Tohoku University, Japan

12:00pm	*LUNCH*

1:30 PM	*Invited Talk 4: Cognitive Radio*

Prof. Bora Nikolic
University of California, Berkeley, U.S.A.

3:00PM	**A Practical Step Forward Toward Software-Defined Radio Transmitters**

*Essam Atalla , Imran Bashir , Poras Balsara , Kamran Kiasaleh , and Robert Bogdan Staszewski**
University of Texas at Dallas, *Texas Instruments Inc., U.S.A.

3:20PM	**Dual-Threshold Voltage Technique for Asynchronous PCFB Linear-Pipelines**

Behnam Ghavami and Hossein Pedram
Amirkabir University of Technology, Tehran, Iran

3:40PM	*BREAK*

2:30PM	***FRIDAY POSTERS***

A Technique to Extend Tuning Range of High Frequency Quadrature VCO
Diptendu Ghosh and Ranjit Gharpurey
University of Texas, Austin, U.S.A.

Top-Down Simulation Methodology of a Mixed-Signal Read Channel Using Standard VHDL
Robert Staszewski
Texas Instruments, Dallas, U.S.A.

Modeling of an Electronic Noise and Media in a Magnetic Recording Read Channel Using VHDL
Robert Staszewski
Texas Instruments, Dallas, U.S.A.

PID-Controlled PLL for Fast Frequency-Hopped Systems
Nil Tarim and Hayri Uyanik
Istanbul Technical University, Turkey

Optimum Designing of the Cascaded Digital Filters in Wide-band Wireless Transmitters Using Genetic Algorithm
Viral Parikh , Sankalp Modi , and Poras Balsara
University of Texas at Dallas, U.S.A.

A 400MHz-2.4Ghz Radiation-Tolerant Self-Biased Phase-Locked Loop
P. Zhu, P. Gui, W. Chen, A.Xiang, D. Gong, T. Liu, Y. Fan, H. Huang*, M. Morgan**
Southern Methodist University, *Texas Instruments Inc. U.S.A.

A Speed and Accuracy-Enhanced On-Chip Current Sensor with Local Shunt Feedback
Hoi Lee and Mengmeng Du
University of Texas at Dallas, U.S.A.

4:30 PM *Poster Removal*

Table of Contents

Keynote and Tutorials

Si Technology Roadmap for Ubiquitous Computing, Sensing, and Perception
Dennis Buss 1

Making Better Use of Time in Mixed-Signal Circuits
Michael Perrott 9

Low Power Reconfigurable Analog-to-Digital Converters
Kush Gulati 17

Verification of Complex Analog Integrated Circuits
Ken Kundert 25

Towards Efficient Spectrum Sharing
Bora Nikolic 33

Presentations

Linearization of Highly-Efficient Monolithic Class E SiGe Power Amplifiers with Envelope-Tracking (ET)
Donald Y.C. Lie, J.D. Popp, F. Wang, D. Kimball, and L.E. Larson 39

Digital FIR Filter Optimization Using Toggle-Based Power Estimation Tools
Albina Cristian, Günther Hackl 43

Analysis of Third-order Intermodulation in Receiver Down-converter with Multiband Feedback
Junghwan Han, Ranjit Gharpurey 47

Enhancement of Coexistence Performance in a DRP Based Multi-Radio Environment of a Mobile Phone
Yossi Tsfati, Oren Eliezer, Ran Katz, Yaniv Tzoref, Ofer Friedman 51

Waveform Analysis and Delay Prediction in Simultaneously Switching CMOS Gate Driven Inductively and Capacitively Coupled On-Chip Interconnects
B.K. Kaushik, S. Sarkar, R.P. Agarwal, R.C. Joshi 55

Design of a Multi-Context FPVLSI based on an Asynchronous Bit-Serial Architecture
Waidyasooriya Hasitha Muthumala, Masanori Hariyama, Michitaka Kameyama 59

A Practical Step Forward Toward Software-Defined Radio Transmitters
Essam Atalla , Imran Bashir, Poras Balsara , Kamran Kiasaleh , Robert Bogdan Staszewski 63

Dual-Threshold Voltage Technique for Asynchronous PCFB Linear-Pipelines
Behnam Ghavami, Hossein Pedram 67

Posters

Layout Parasitic Interconnections effects on High Frequency Circuits
Albina Cristian, Günther Hackl   71

"Flying-Adder" PLL Based Synchronization Mechanism for Data Packet Transport
Liming Xiu, Steve Clynes, Srikanth Gurrapu, Towfique Haider, Feng Ying,
Wahed Mohammed   75

System-On-Chip Power Consumption Refinement and Analysis
David Feinstein, Mitch Thornton, Faith Kocan   81

A High-Performance Multi-Match Priority Encoder for TCAM-Based Packet Classifiers
Mehrdad Nourani, Faezipour Miad   85

An Efficient Implementation of Scalable Architecture for Discrete Wavelet Transform on FPGA
Hassan,RABAH, Michael Guarisco, Xun Zhang, Serge Weber   89

A New Fast Slew Buffering Algorithm without Input Slew Assumptions
Shiyan Hu, Jiang Hu   93

Closed Form Equations for Inter-Modulation Distortion Parameters in WCDMA Receiver Validated Through Measurements
Mohammed Saif Khan, Naveen K. Yanduru   97

A CMOS Wideband LNA Using Multiple Phase Matched Frequency Staggered Resonators
Diptendu Ghosh, Ranjit Gharpurey   101

A Technique to Extend Tuning Range of High Frequency Quadrature VCO
Diptendu Ghosh, Ranjit Gharpurey   105

Top-Down Simulation Methodology of a Mixed-Signal Read Channel Using Standard VHDL
Robert Staszewski   109

Modeling of an Electronic Noise and Media in a Magnetic Recording Read Channel Using VHDL
Robert Staszewski   113

PID-Controlled PLL for Fast Frequency-Hopped Systems
Nil Tarim, Hayri Üyanik   117

Optimum Designing of the Cascaded Digital Filters in Wide-band Wireless Transmitters Using Genetic Algorithm
Viral Parikh, Sankalp Modi, Poras Balsara   121

A 400MHz-2.4Ghz Radiation-Tolerant Self-Biased Phase-Locked Loop
P. Zhu, P. Gui, W. Chen, A. Xiang, D. Gong, T. Liu, Y. Fan, H. Huang, M. Morgan   125

A Speed- and Accuracy-Enhanced On-Chip Current Sensor with Local Shunt Feedback
Hoi Lee, Mengmeng Du   129

Si TECHNOLOGY ROADMAP FOR UBIQUITOUS COMPUTING, SENSING, AND PERCEPTION

Dr Dennis Buss
Texas Instruments Inc

SINGLE CHIP PHONE

AGENDA

- Moore's Law Scaling
- Design for Low Power
- SOC Integration of Analog/RF Functions
- Digital Radio Processor
- Integrated MEMS
- Conclusion

Semi-Conductor Scaling

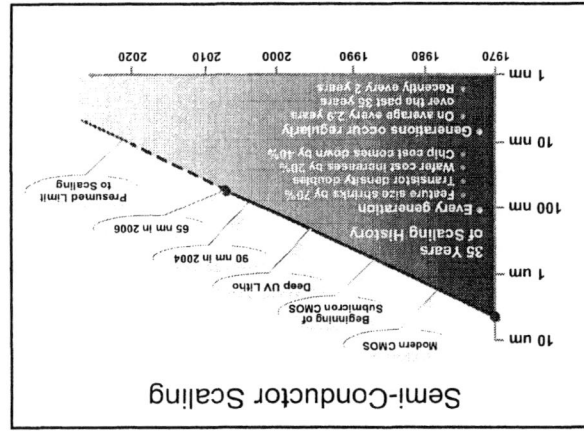

GSM Digital Baseband Evolution

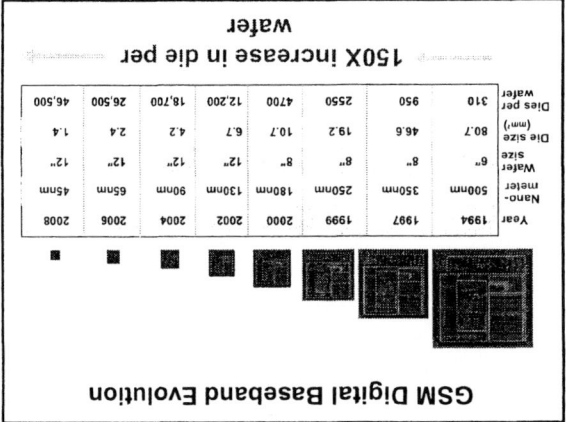

150X increase in die per wafer

Year	1994	1997	1999	2000	2002	2004	2006	2008
Nano-meter	500nm	350nm	250nm	180nm	130nm	90nm	65nm	45nm
Wafer size	6"	8"	8"	12"	12"	12"	12"	12"
Die size (mm)	80.7	46.6	19.2	10.7	6.7	4.2	2.4	1.4
Dies per wafer	310	950	2550	4700	12,200	18,700	26,500	46,500

First 65 nm Product: DBB chip

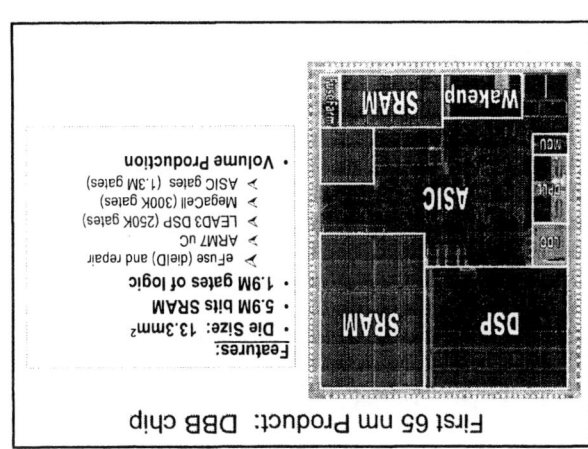

Features:
- **Die Size: 13.3mm²**
- **5.9M bits SRAM**
- **1.9M gates of logic**
  - ➤ eFuse (dieID) and repair
  - ➤ ARM7 uC
  - ➤ LEAD3 DSP (250K gates)
  - ➤ MegaCell (300K gates)
  - ➤ ASIC gates (1.3M gates)
- Volume Production

---

65 nm Features

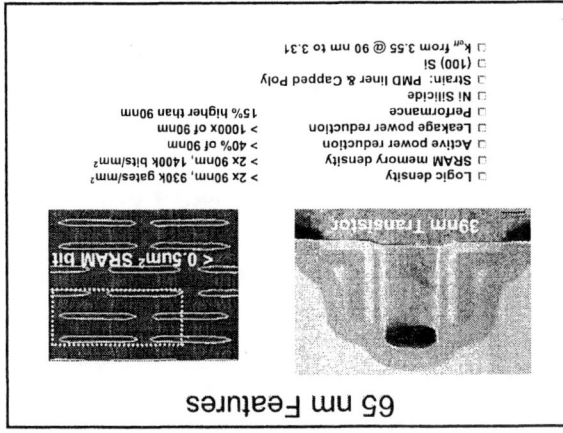

< 0.5µm² SRAM bit

39nm Transistor

- □ Logic density   > 2x 90nm, 930K gates/mm²
- □ SRAM memory density   > 2x 90nm, 1400K bits/mm²
- □ Active power reduction   > 40% of 90nm
- □ Leakage power reduction   > 1000X of 90nm
- □ Performance   15% higher than 90nm
- □ Ni Silicide
- □ Strain: PMD liner & Capped Poly
- □ (100) Si
- □ K_eff from 3.55 @ 90 nm to 3.31

Power Domain Partitioning

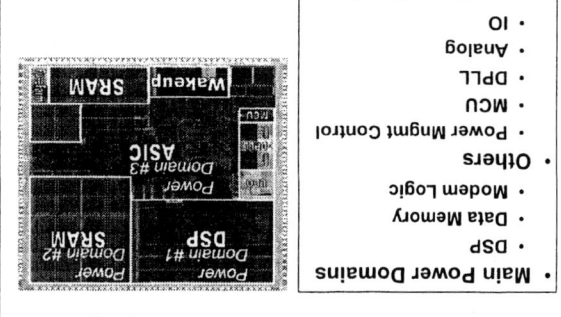

Main Power Domains
- DSP
- Data Memory
- Modem Logic
- Others
- Power Mngmt Control
- MCU
- DPLL
- Analog
- IO

Deep Submicron Processes Demand Enhanced Power Management

Talk Time: Pwr Active = CV²F + Leakage
- C: Decrease/node, offset by complexity
- F: Increases/node
- Leakage: increases/node, temp.

Standby Time: Pwr Idle = Leakage
- Leakage: increases/node, temp

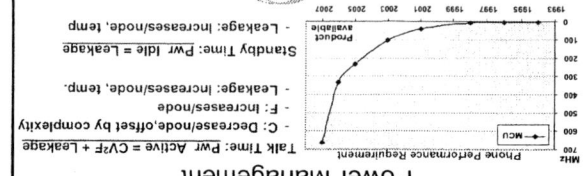
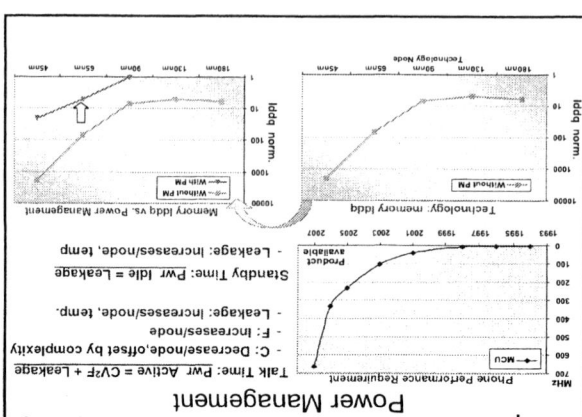

SINGLE CHIP PHONE

AGENDA

- Moore's Law Scaling
▶ Design for Low Power
- SOC Integration of Analog Functions
- Digital Radio Processor
- Integrated MEMS
- Conclusion

TI CMOS Roadmap
Issues/Trends

- Relentless focus on power reduction.
- Design for Manufacturing Variations
- Analog/RF SOC Integration
- Co-Development of Process, Design techniques and Architecture
- CMOS processes customized to the application

STRUCTURED LAYOUT (45 nm)

ACTIVE
- Max Xstor width change within
- Poly not required to overlap contact
- GHOST Poly
- Vertical poly gates only

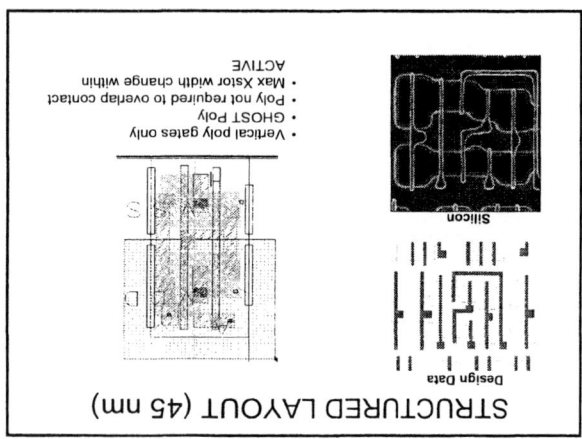

Lithography

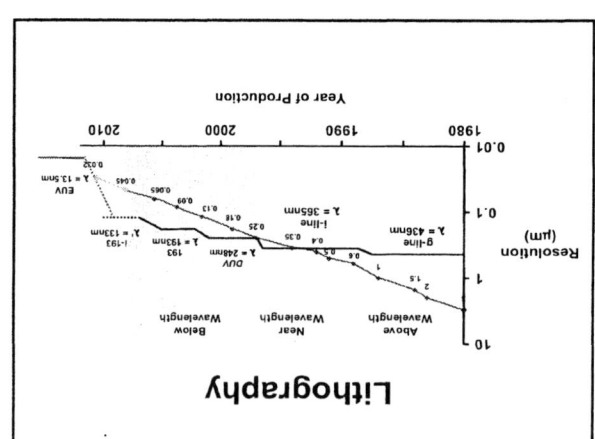

SINGLE CHIP PHONE

AGENDA

- Moore's Law Scaling
- Design for Low Power
- ◀ SOC Integration of Analog Functions
- Digital Radio Processor
- Integrated MEMS
- Conclusion

Typical Cell-Phone Block Diagram

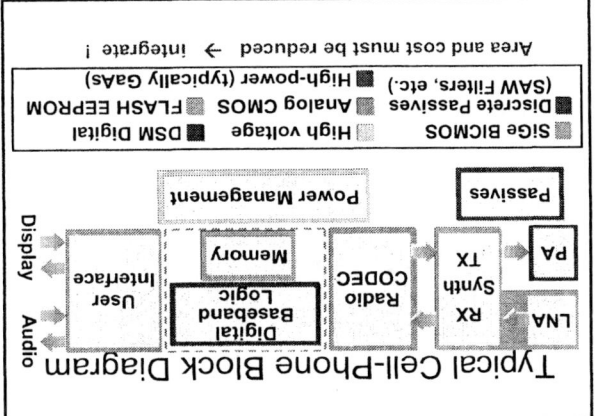

Legend:
- SiGe BiCMOS | High voltage | DSM Digital
- Discrete Passives | Analog CMOS | FLASH EEPROM
- (SAW Filters, etc.) | High-power (typically GaAs)

Area and cost must be reduced → integrate !

Memory Retention

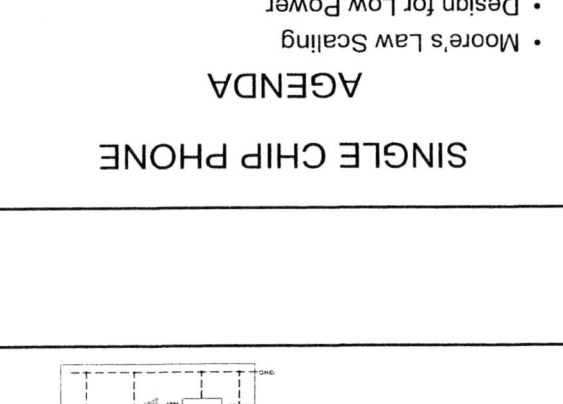

- **Active Mode**
  - Periphery On
  - VSSM = 0V
  - VDD = V Nominal
- **Retention**
  - Periphery Off
  - VSSM ~= 0.5V
  - VDD ~=1.0V

Silicon Measurements

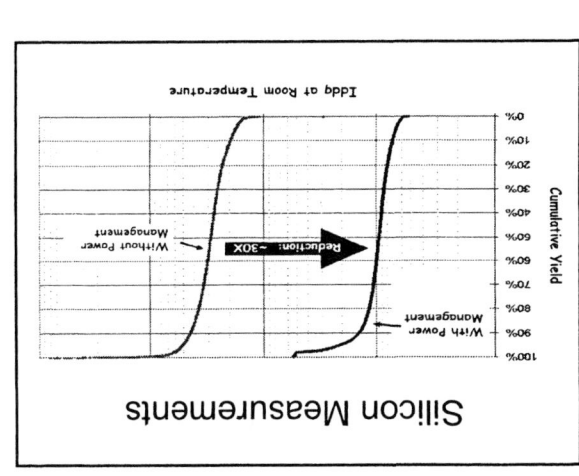

Idd at Room Temperature

Without Power Management — Reduction: ~30X — With Power Management

---

Why Single-Chip Phone?

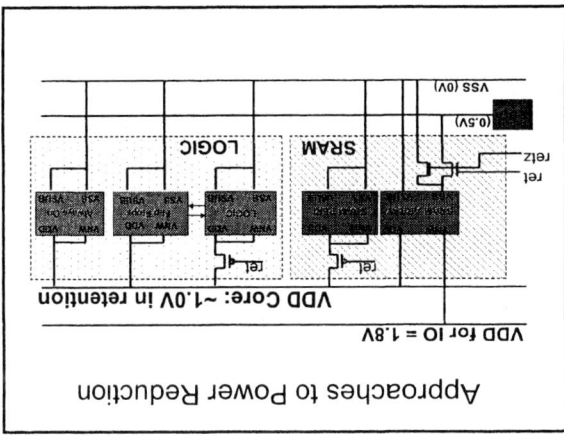

- "Integration is like gravity"
  - Already happened in hard-disk drives, ADSL, etc
  - Not a single example of reversal
- "$20 phones"
- Large untapped market in India and China
- More "real estate" space for advanced features
- Better reliability
  - Today, more than half of the total components on a board are analog RF components
- Longer talk time

Approaches to Power Reduction

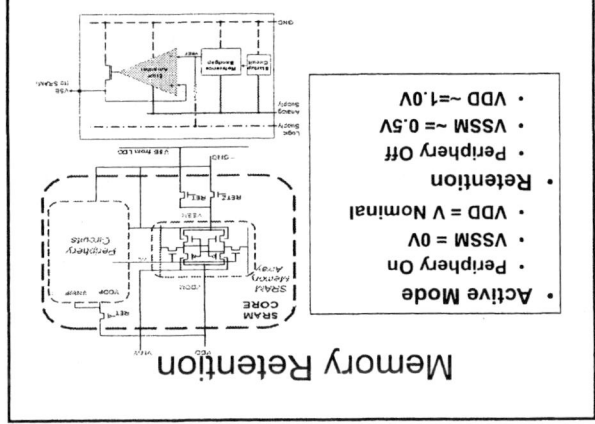

VDD for IO = 1.8V

VDD Core: ~1.0V in retention

LOGIC   SRAM   rel1   rel2 (0.5V)   VSS (0V)

SINGLE CHIP PHONE

AGENDA

- Moore's Law Scaling
- Design for Low Power
- SOC Integration of Analog Functions
- ▶ Digital Radio Processor
- Integrated MEMS
- Conclusion

Conventional Transceivers

- RF transmitters in commercial wireless applications are traditionally based on charge-pump PLL's and IQ upconversion mixers
- RF receivers use continuous-time mixing, filtering and amplification
- Design flow and circuit techniques are analog intensive
- Technology incompatible with modern digital processors
  – Low-voltage deep-submicron CMOS

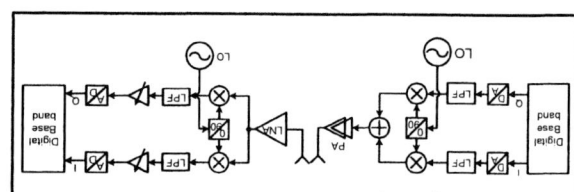
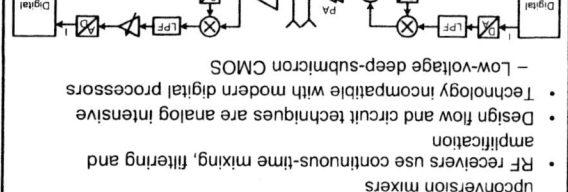

DRP RF Architecture

0.2, 1.25, 2.5.5, 10, 15, 20MHz BW

Transmission bands
450, 800, 900, 1800, 1900, 2100, 2500, 3400MHz

GMSK, QPSK, 8-PSK, 16QAM, 64QAM Modulation

TDMA, FDMA, CDMA, OFDMA, FDMA schemes

Looks like SDR!!

RX: RF in → LNTA → Current sampler → Discrete time → A/D → Digital Logic → RX data

TX: TRXX Combiner, DPA → RF out, TDC, Lo clock, ΣΔ, DCO, Digital Logic, Amplitude Regulation, Channel, TX data

Power Management (PM)

RF Built-in Self Test (RFBiST)

Digital Baseband and Application Processor

DRP/SoC Proven in Many Products

- 1/2 the silicon
- 1/2 the power
- 1/2 the board area

More to come..

"LoCosto" GSM/GPRS · "Hollywood" mDTV · WiLink™ Wi-Fi · NaviLink™ A-GPS · BlueLink™ Bluetooth

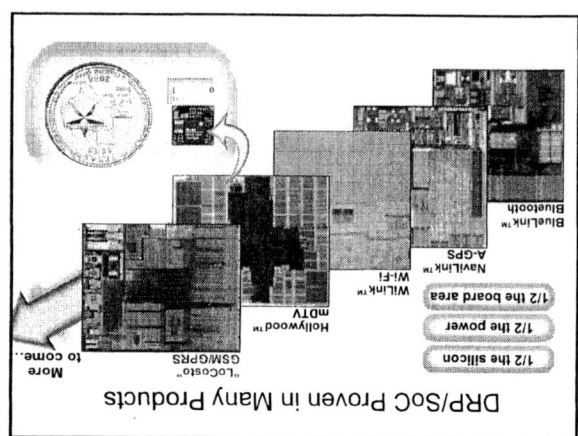

Single-Chip GSM Radio

- 90 nm CMOS
- All-digital PLL
- All-digital TX
- Digitally-intensive RX
- w/o
  – 2-W PA
  – Battery management

RX: LNA → Current sampler → Discrete time → A/D → Digital logic

TX: DPA, DCO, ΣΔ, ΣΔ, AM, DCXO, xo, Front-end Module

SRAM, Digital Baseband Processor

RF Built-in Self Test · Power Management · Battery Management · V_bat

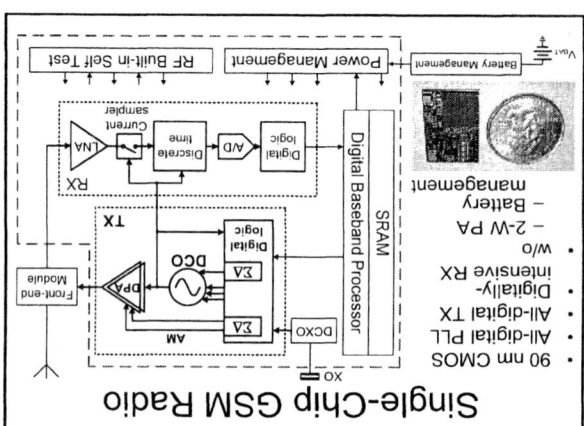

What about SiP Integration?

- Monolithic integration of DRAM would result in significant cost increase due to the need for additional mask levels.
- Memory modules are highly reusable so modularity makes sense.
- No yield impact issue due to in-package integration of memory.

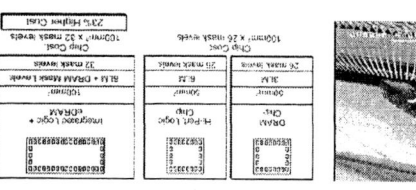

SINGLE CHIP PHONE

AGENDA

- Moore's Law Scaling
- Design for Low Power
- SOC Integration of Analog Functions
- Digital Radio Processor
- ▶ Integrated MEMS
- Conclusion

Digital Radios Offer Many Benefits
Why Digital?

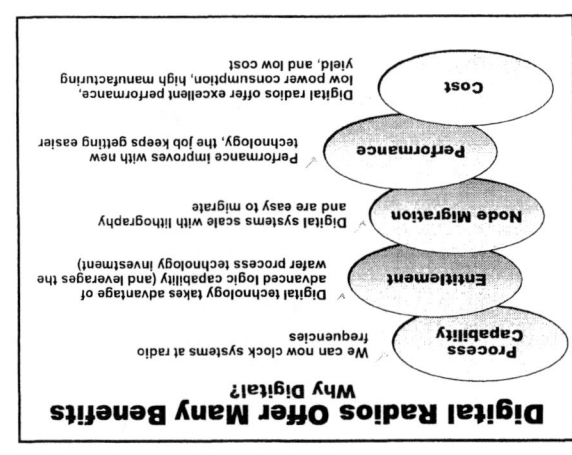

Process Capability	✓ We can now clock systems at radio frequencies
Entitlement	✓ Digital technology takes advantage of advanced logic capability (and leverages the wafer process technology investment)
Node Migration	Digital systems scale with lithography and are easy to migrate
Performance	Performance improves with new technology, the job keeps getting easier
Cost	Digital radios offer excellent performance, low power consumption, high manufacturing yield, and low cost

---

Deep-Submicron CMOS Rules

- Exploit:
  - Fast switching characteristics of MOS transistors
  - Small device geometries and precise device matching
  - High density of digital logic: 250 kgates/mm² in 90-nm CMOS
  - High density of SRAM memory: 1 Mbits / mm² in 90-nm CMOS
- Avoid:
  - Biasing currents for analog circuits
  - Reliance on voltage resolution
  - Nonstandard devices not needed for memory and digital logic

---

SoC Drives Cost Reduction

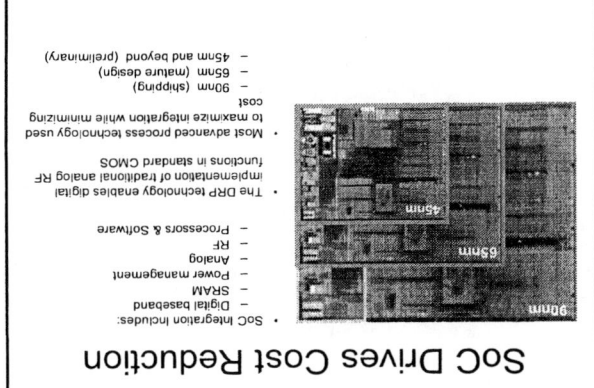

- SoC Integration includes:
  - Digital baseband
  - SRAM
  - Power management
  - Analog
  - RF
  - Processors & Software
- The DRP technology enables digital implementation of traditional analog RF functions in standard CMOS
- Most advanced process technology used to maximize integration while minimizing cost
  - 90nm (shipping)
  - 65nm (mature design)
  - 45nm and beyond (preliminary)

All-Digital PLL vs. Charge-pump PLL

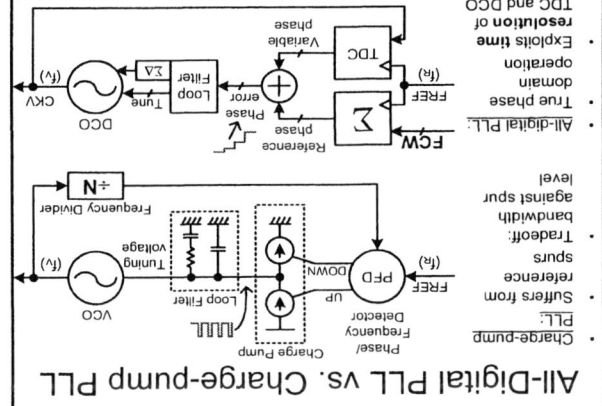

- Charge-pump PLL:
  - Suffers from reference spurs
  - Tradeoff: bandwidth against spur level
- All-digital PLL:
  - True phase domain operation
  - Exploits time resolution of TDC and DCO

New Paradigm

In a deep-submicron CMOS process, time-domain resolution of a digital signal edge transition is superior to voltage resolution of an analog signal

How the DMD Works

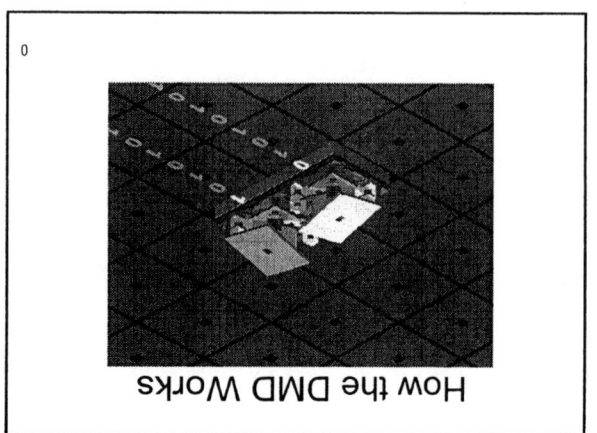

How the DMD Works

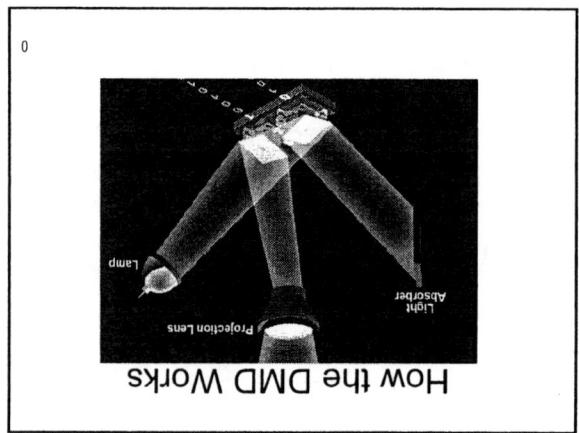

How the DMD Works

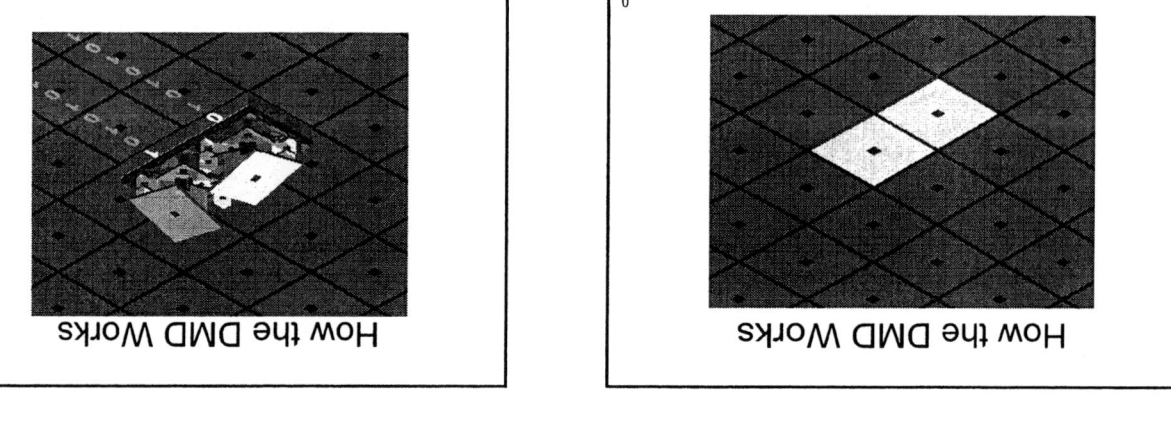

How the DMD Works

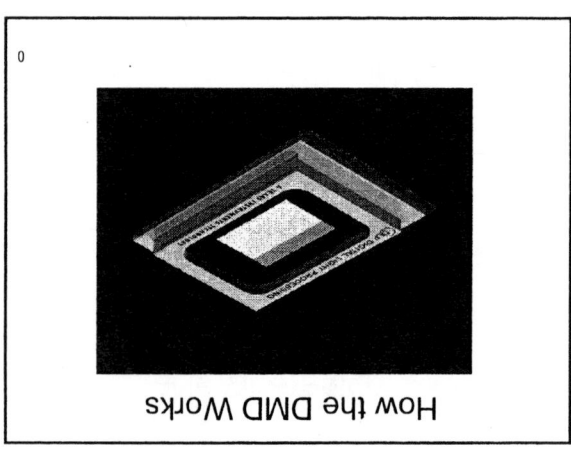

MEMS Integration

- MEMS Integration will also enable products for ubiquitous computing, sensing and perception
  - accelerometers
  - pressure sensors
  - rate gyros
  - Integrated microphones
  - Resonators
  - RF switches and tuneable capacitors
  - optical switches and phase modulators,
  - micro-fluidic pumps and valves
  - Displays
- The Digital Mirror Device (DMD) is an example of integrated MEMS

Technology in the Next Decade

Moore's Law is predicted to stagnate toward the end of the next decade …

… but SOC integration has the potential to continue IC cost reduction and to perpetuate growth of products for ubiquitous computing, perception & sensing.

Technology in the Next Decade

Moore's Law is predicted to stagnate toward the end of the next decade …

SINGLE CHIP PHONE

AGENDA

- Moore's Law Scaling
- Design for Low Power
- SOC Integration of Analog Functions
- Digital Radio Processor
- Integrated MEMS
- ◀ Conclusion

MEMS Integration

Today
•HDTV
•Front Projection Products
•Large Screen Movie Theaters

Tomorrow
•Projection displays for cell phones and PDAs
•3D imaging for medical

Making Better Use of Time in Mixed Signal Circuits

Michael H. Perrott
Massachusetts Institute of Technology
November 2007

|||iī CICS Center for Integrated Circuits and Systems

Massachusetts Institute of Technology

Copyright © 2007 by Michael H. Perrott

All rights reserved.

Motivation

- The world is changing for analog designers
  – Future processes are offering degraded analog device characteristics (lower g_m/r_o, lower supply voltage, etc.)
- There is pressing need for new mixed-signal circuit architectures
  – Leverage digital circuits to perform analog processing
- An interesting observation: representation of signals as transition times allows digital circuits to perform analog signal processing
  – Phase-locked loop circuits have been the classical benefactor of this observation

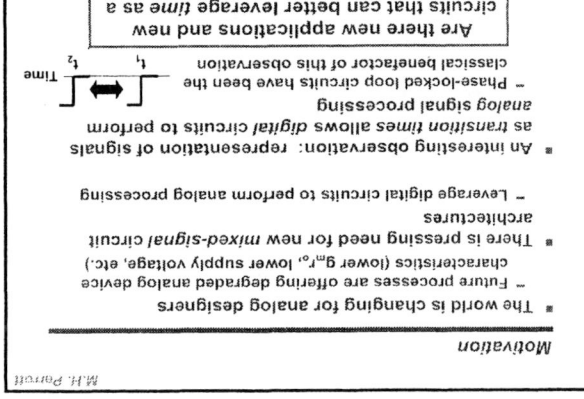

Are there new applications and new circuits that can better leverage time as a signal domain?

...Just Enough PLL Background...

What is a Phase-Locked Loop (PLL)?

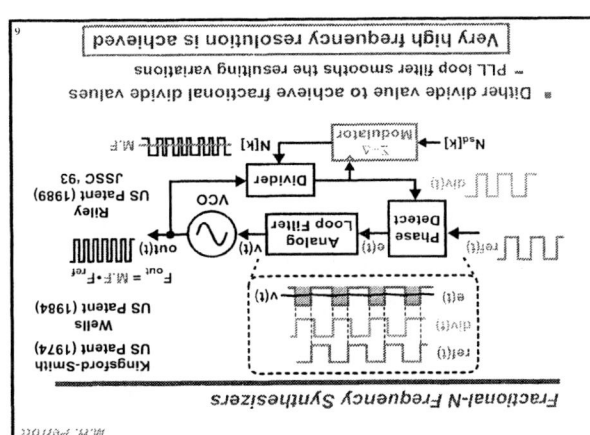

de Bellescize
Onde Electr, 1932

- VCO efficiently provides oscillating waveform with variable frequency
- PLL synchronizes VCO frequency to input reference frequency through feedback
  – Key block is phase detector
  – Realized as *digital gates* that create pulsed signals

Integer-N Frequency Synthesizers

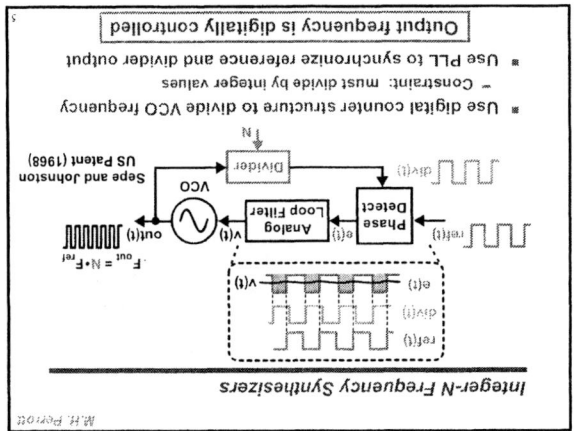

Sepe and Johnson
US Patent (1968)

$F_{out} = N \cdot F_{ref}$

- Use digital counter structure to divide VCO frequency
  – Constraint: must divide by integer values
- Use PLL to synchronize reference and divider output

Output frequency is digitally controlled

Fractional-N Frequency Synthesizers

Kingsford-Smith US Patent (1974)
Wells US Patent (1984)
Riley US Patent (1989) JSSC '93

$F_{out} = M.F \cdot F_{ref}$

$N_{sd}[k]$, $N[k]$, $M.F$, Σ-Δ Modulator

- Dither divide value to achieve fractional divide values
  – PLL loop filter smooths the resulting variations

Very high frequency resolution is achieved

The Issue of Quantization Noise

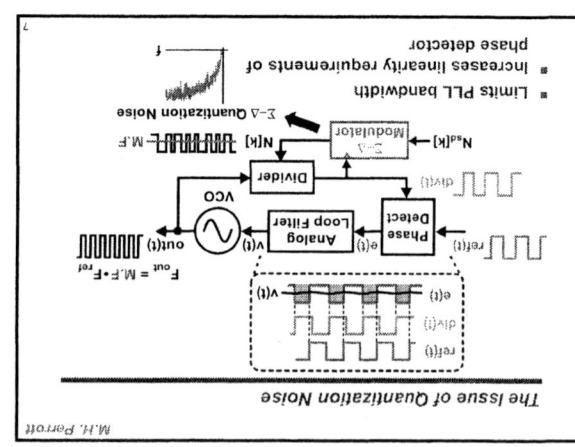

$F_{out} = M.F_{ref}$

Σ-Δ Quantization Noise

- Limits PLL bandwidth
- Increases linearity requirements of phase detector

Striving for a Better PLL Implementation

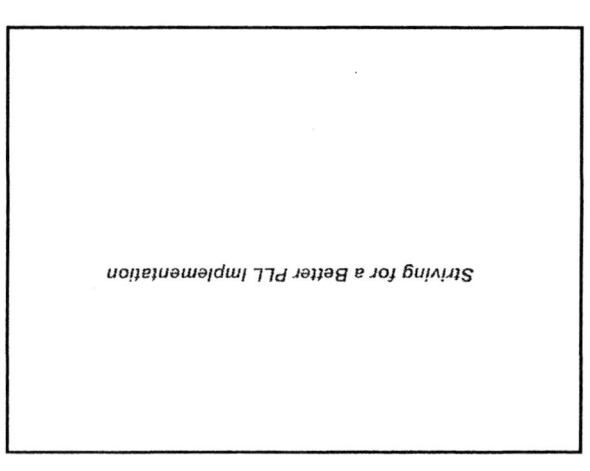

Analog Phase Detection

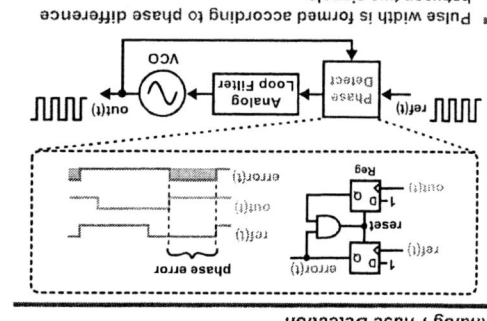

phase error

- Pulse width is formed according to phase difference between two signals
- Average of pulsed waveform is applied to VCO input

Tradeoffs of Analog Approach

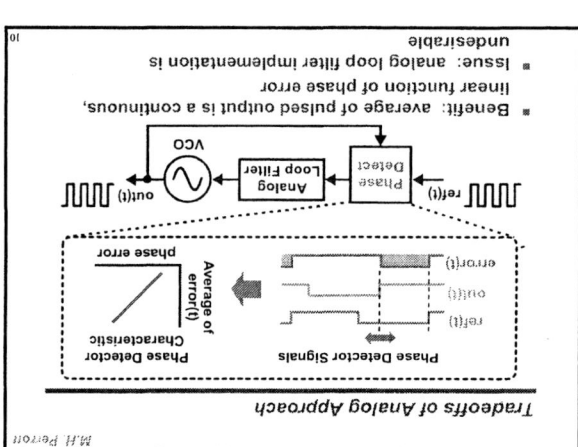

Phase Detector Signals — Average of error(t) — Phase Detector Characteristic — phase error

- Benefit: average of pulsed output is a continuous, linear function of phase error
- Issue: analog loop filter implementation is undesirable

Going Digital ...

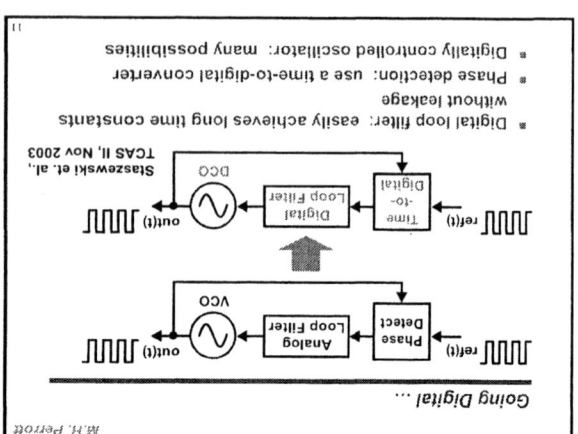

Staszewski et. al., TCAS II, Nov 2003

- Digital loop filter: easily achieves long time constants
- Phase detection: use a time-to-digital converter without leakage
- Digitally controlled oscillator: many possibilities

Classical Time-to-Digital Converter

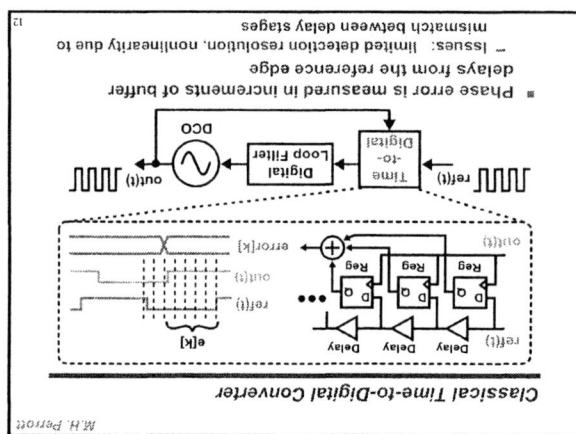

- Phase error is measured in increments of buffer delays from the reference edge
- Issues: limited detection resolution, nonlinearity due to mismatch between delay stages

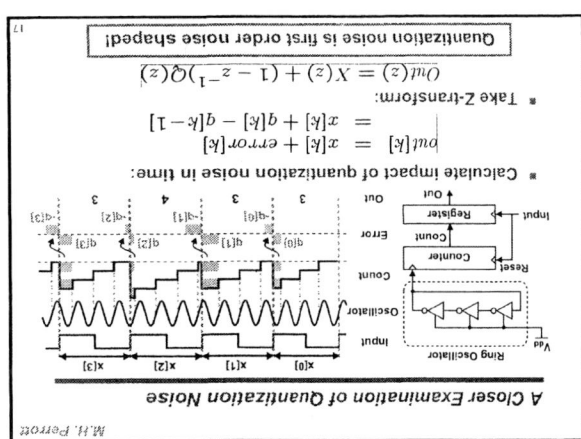

Impact of Limited Resolution and Nonlinearity

- Integer-N PLL: limit cycles
- Fractional-N PLL: undesired fractional spurs

Additional filtering is not effective in fixing these issues

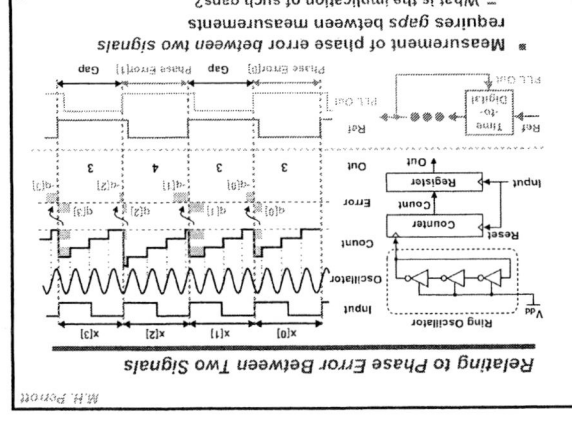

Proposed Approach: A Better Time-to-Digital Converter

Helal, Straayer, Perrott VLSI 2007

- This is a simplified view
  - We will need a few slides to properly explain this ...

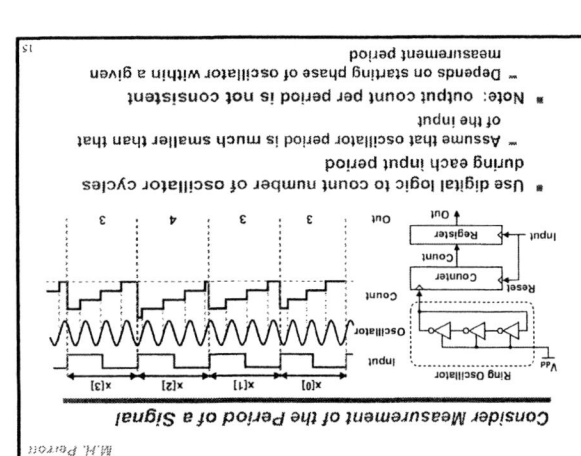

Consider Measurement of the Period of a Signal

- Use digital logic to count number of oscillator cycles during each input period
  - Assume that oscillator period is much smaller than that of the input
- Note: output count per period is not consistent
  - Depends on starting phase of oscillator within a given measurement period

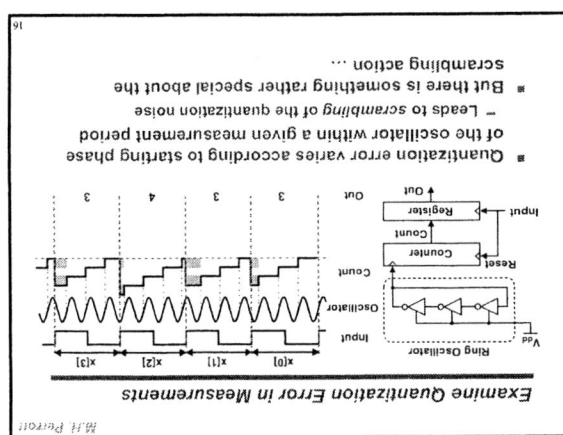

Examine Quantization Error in Measurements

- Quantization error varies according to starting phase of the oscillator within a given measurement period
  - Leads to scrambling of the quantization noise
- But there is something rather special about the scrambling action ...

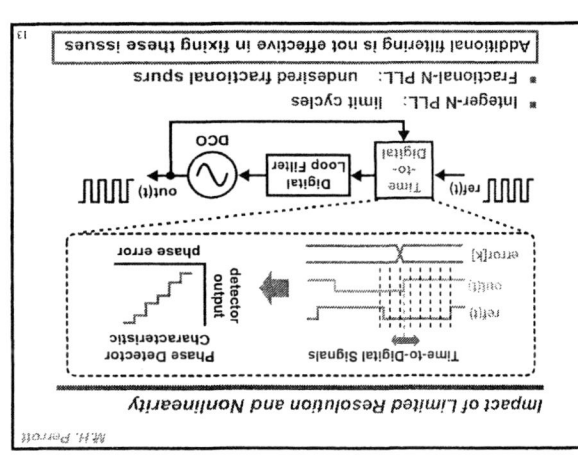

A Closer Examination of Quantization Noise

- Calculate impact of quantization noise in time:

$$out[k] = x[k] + error[k]$$
$$= x[k] + q[k] - q[k-1]$$

- Take Z-transform:

$$Out(z) = X(z) + (1 - z^{-1})Q(z)$$

Quantization noise is first order noise shaped!

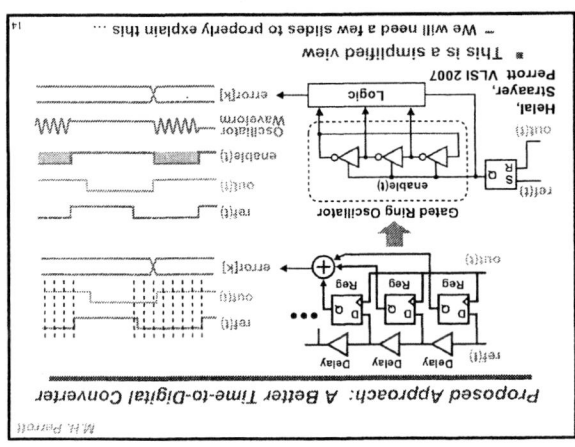

Relating to Phase Error Between Two Signals

- Measurement of phase error between two signals requires gaps between measurements
  - What is the implication of such gaps?

PLL Applications of Proposed GRO Structure

Measured GRO Results Confirm Noise Shaping

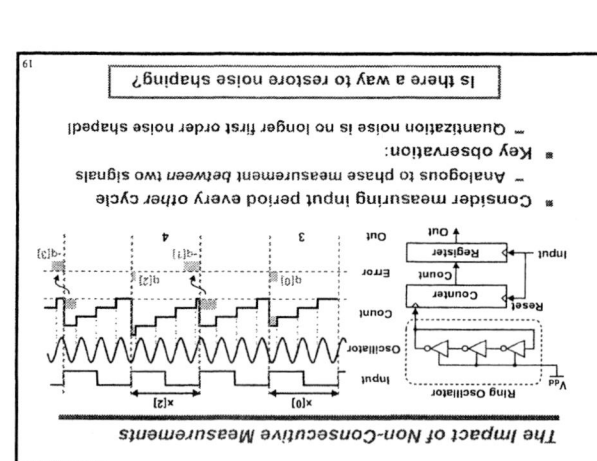

- Input variable delay signal
- Harmonics due to nonlinearity of variable delay
- Noise shaped quant. noise

GRO Prototype

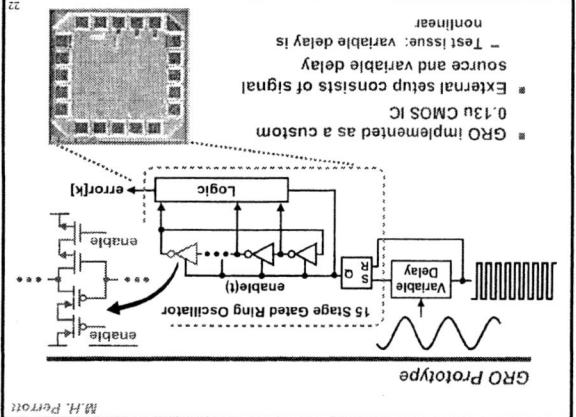

- GRO implemented as a custom 0.13u CMOS IC
- External setup consists of signal source and variable delay
  - Test issue: variable delay is nonlinear

Improve Resolution By Using All Oscillator Phases

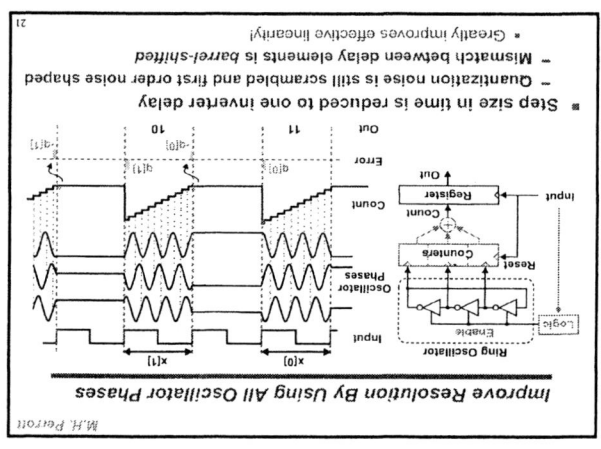

- Step size in time is reduced to one inverter delay
  - Quantization noise is still scrambled and first order noise shaped
  - Mismatch between delay elements is barrel-shifted
    - Greatly improves effective linearity!

Proposed Approach: Gate the Oscillator

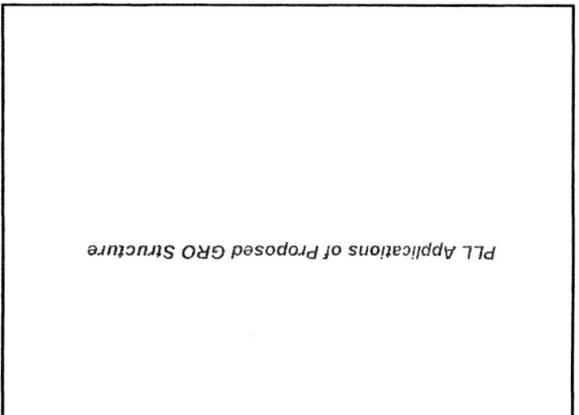

- Key requirements of approach:
  - Turn off oscillator at conclusion of each measurement
  - Be sure that state information within oscillator is preserved
- Reasonably straightforward with CMOS ring oscillators

First order noise shaping is restored!

The Impact of Non-Consecutive Measurements

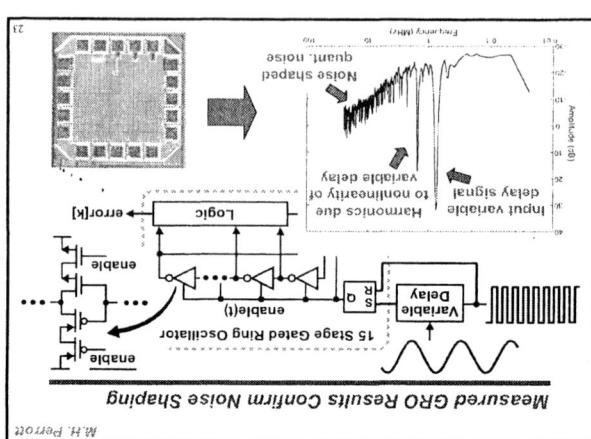

- Consider measuring input period every other cycle
  - Analogous to phase measurement between two signals
- Key observation:
  - Quantization noise is no longer first order noise shaped!

Is there a way to restore noise shaping?

Multiplying DLL Concept

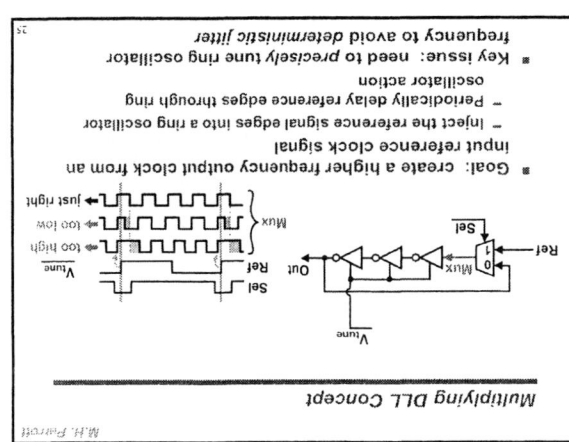

- Goal: create a higher frequency output clock from an input reference clock signal
  - Inject the reference signal edges into a ring oscillator
  - Periodically delay reference edges through ring oscillator action
- Key issue: need to precisely tune ring oscillator frequency to avoid deterministic jitter

---

The Benefit of the MDLL Approach

Ye, Jansson, Galton, JSSC, Dec. 2002

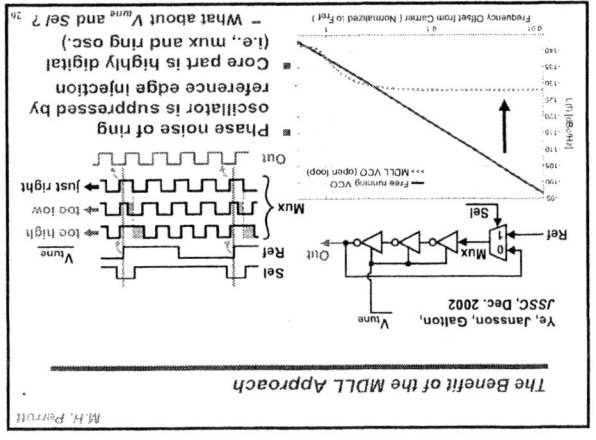

- Phase noise of ring oscillator is suppressed by reference edge injection
- Core part is highly digital (i.e., mux and ring osc.)
  - What about V_{tune} and Sel ?

Let us first look at the classical implementation ...

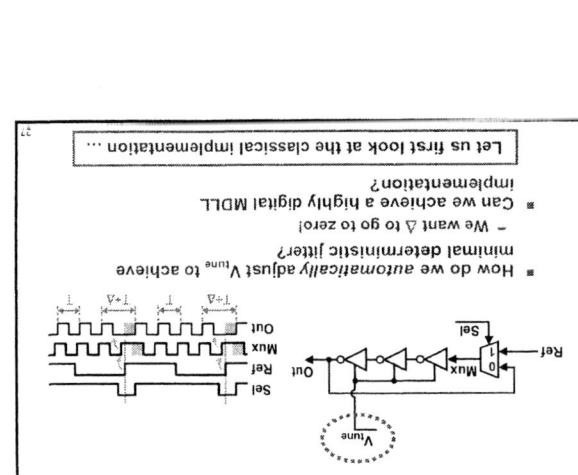

- Can we achieve a highly digital MDLL implementation?
  - We want Δ to go to zero!
- How do we *automatically* adjust V_{tune} to achieve minimal deterministic jitter?

Classical Analog Approach to Adjusting V_{tune}

Farjad-Rad et. al., JSSC, Dec. 2002.

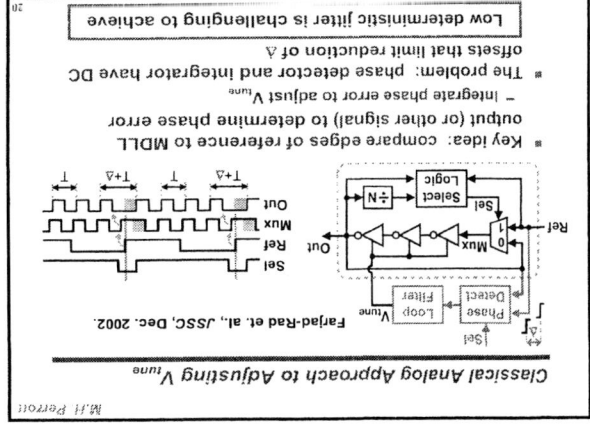

- Key idea: compare edges of reference to MDLL output (or other signal) to determine phase error
  - Integrate phase error to adjust V_{tune}
- The problem: phase detector and integrator have DC offsets that limit reduction of Δ

Low deterministic jitter is challenging to achieve

Proposed Approach

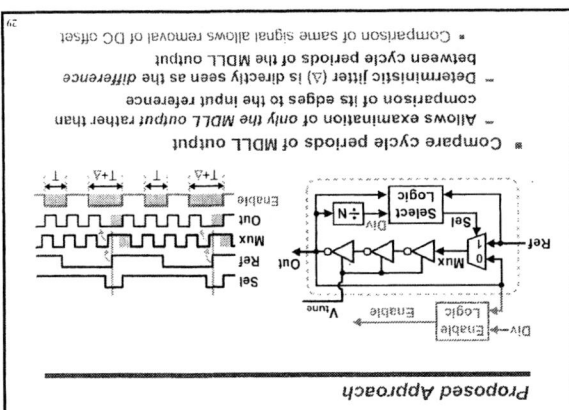

- Compare cycle periods of MDLL output
  - Allows examination of only the MDLL output rather than comparison of its edges to the input reference
  - Deterministic jitter (Δ) is directly seen as the *difference* between cycle periods of the MDLL output
- Comparison of same signal allows removal of DC offset

We Can Achieve a Highly Digital Implementation!

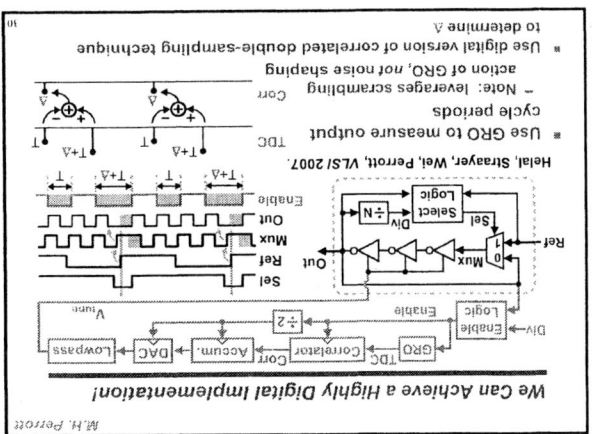

Helal, Straayer, Wei, Perrott, VLSI 2007.

- Use GRO to measure output cycle periods
  - Note: leverages scrambling action of GRO, *not* noise shaping
- Use digital version of correlated double-sampling technique to determine Δ

A First Step Toward Modeling

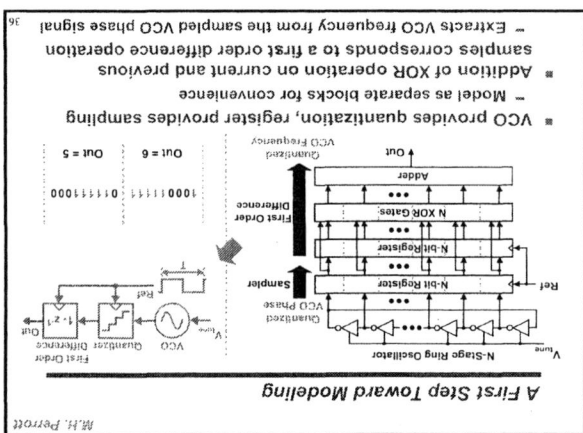

- VCO provides quantization, register provides sampling
  - Model as separate blocks for convenience
- Addition of XOR operation on current and previous samples corresponds to a first order difference operation
  - Extracts VCO frequency from the sampled VCO phase signal

A Better Implementation for High Speed Conversion

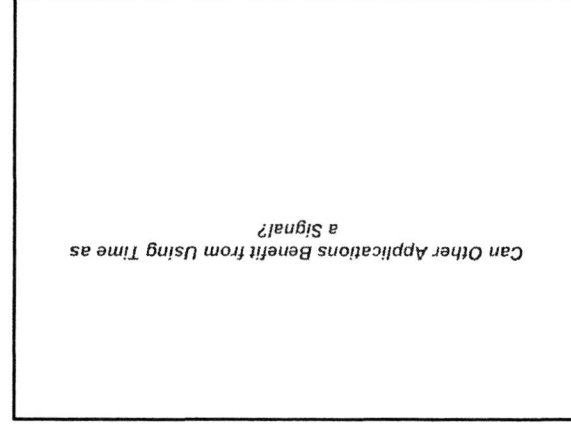

Example: Progression of 9-Stage Ring Oscillator Values

- Assume a high Ref clock frequency (i.e., 1 GHz)
- Increase number of stages, N, such that transitions never cycle through any stage more than once per Ref clock period
- Use registers and XOR gates to determine transition count
  - Avoidance of reset action improves operating speed

Using Time for Analog-to-Digital Conversion

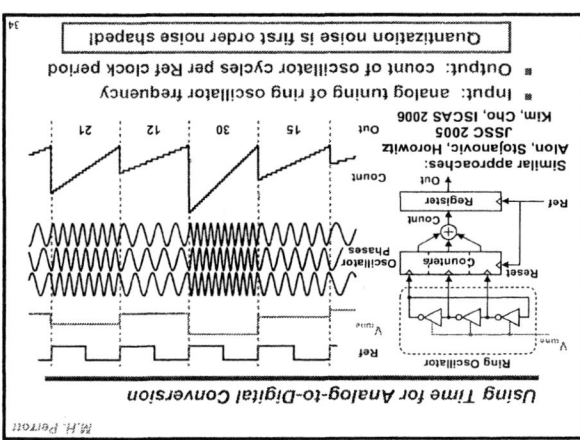

Similar approaches:
Naik, Stojanovic, Horowitz
JSSC 2005
Kim, Cho, ISCAS 2006

- Input: analog tuning of ring oscillator frequency
- Output: count of oscillator cycles per Ref clock period
- Quantization noise is first order noise shaped!

Can Other Applications Benefit from Using Time as a Signal?

Measured Jitter of Prototype

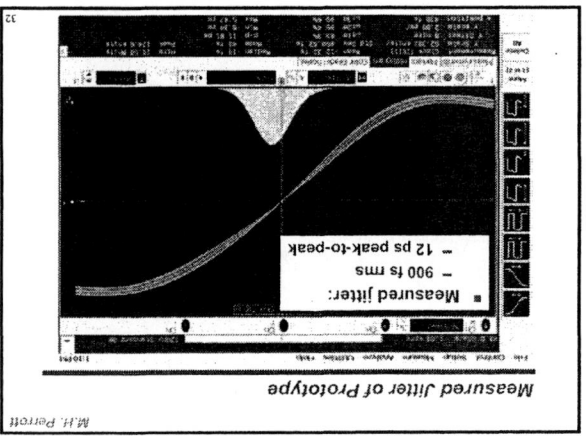

- Measured jitter:
  - 900 fs rms
  - 12 ps peak-to-peak

Our Prototype

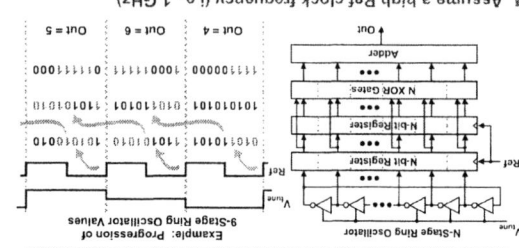

- Two custom 0.13u CMOS ICs
  - GRO and core MDLL structures
- FPGA
  - Correlated double-sampling and accumulation operations
- Discrete 14-bit DAC and RC lowpass (5.3 Mhz pole)
  - 8-bit to 10-bit DAC is adequate (RC pole at 3 MHz)

Key Insight: Quantizer Acts as a Barrel-Shifter

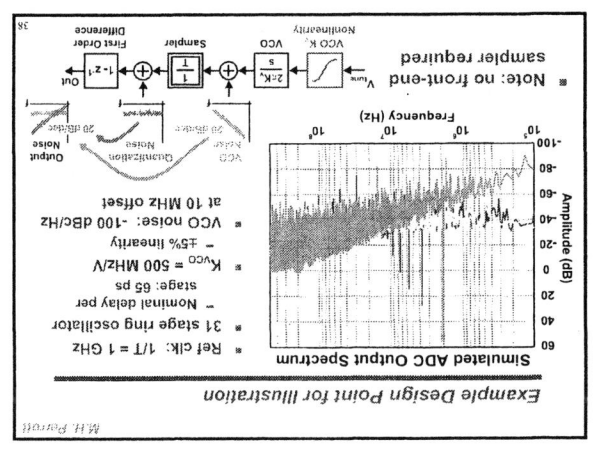

A Closer Look at the DAC Implementation

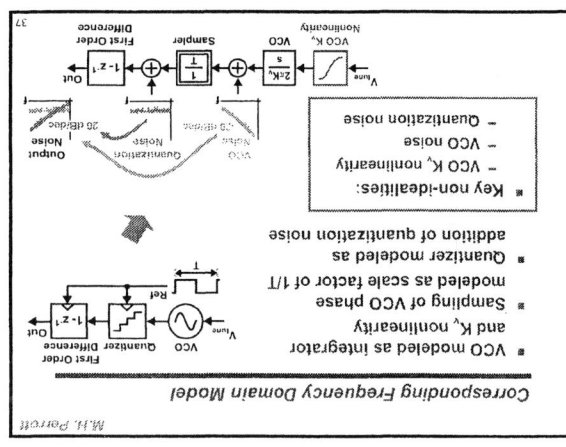

Reducing the Impact of Nonlinearity using Feedback

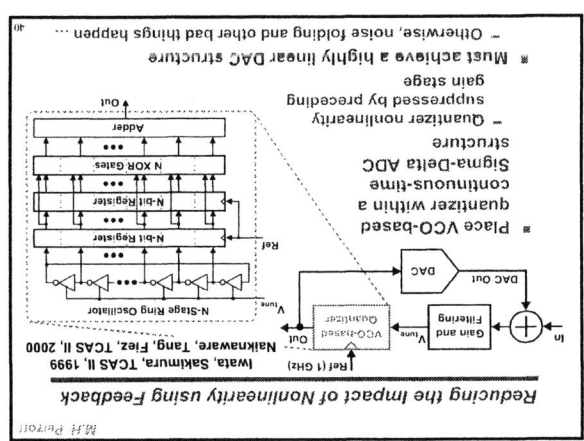

SNR/SNDR Calculations with 20 MHz Bandwidth

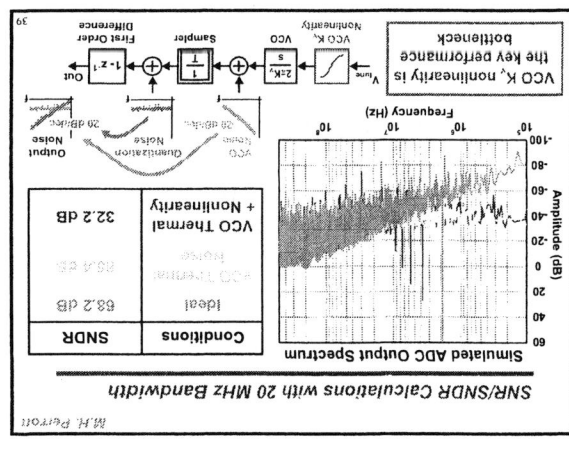

Example Design Point for Illustration

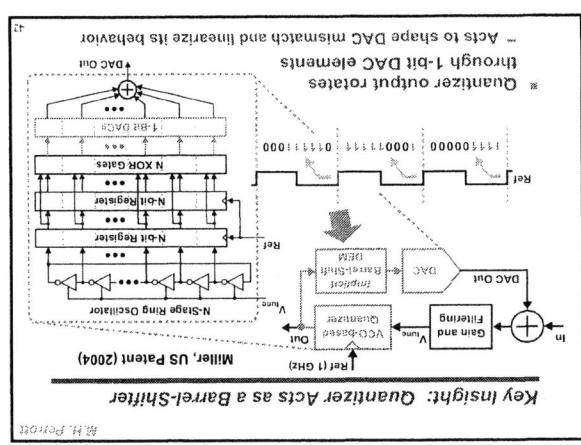

Corresponding Frequency Domain Model

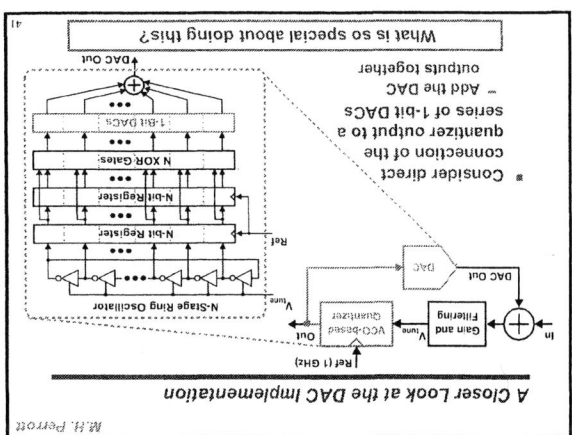

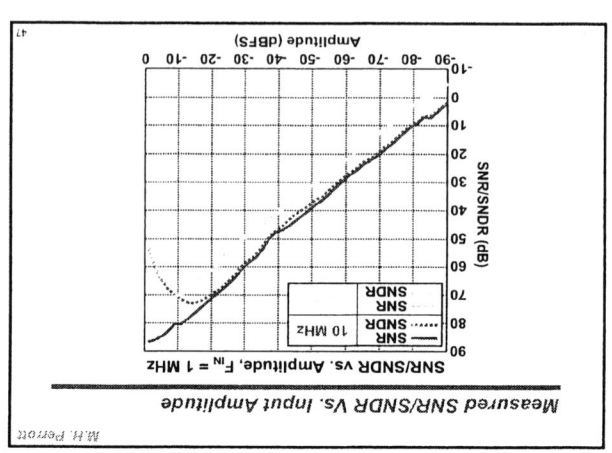

A Geometric View of the VCO Quantizer/DEM and DAC

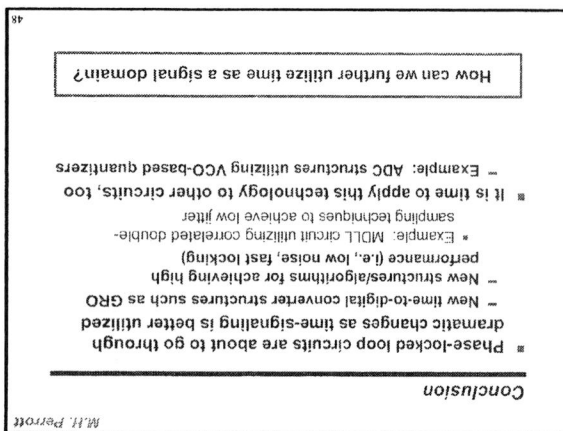

Our Prototype

- Corresponds to classic second order Sigma-Delta Candy structure (Candy, *Trans. On Comm.*, Mar 1985)
- However:
  - Two integrators achieved with only one op-amp
  - *Third order* noise shaping is achieved!
- VCO-based quantizer adds an extra order of noise shaping

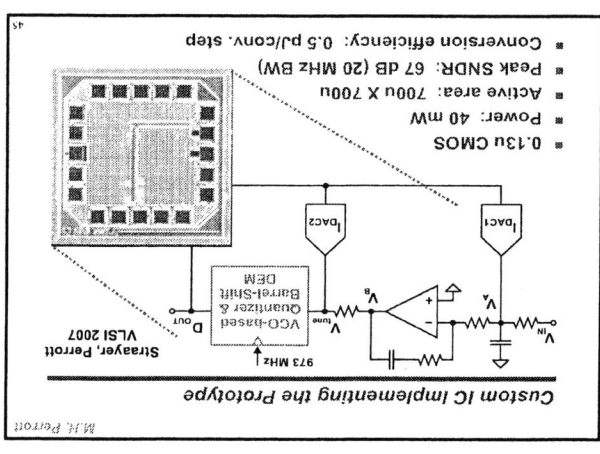

Custom IC Implementing the Prototype

Straayer, Perrott
VLSI 2007

- 0.13u CMOS
- Power: 40 mW
- Active area: 700u X 700u
- Peak SNDR: 67 dB (20 MHz BW)
- Conversion efficiency: 0.5 pJ/conv. step

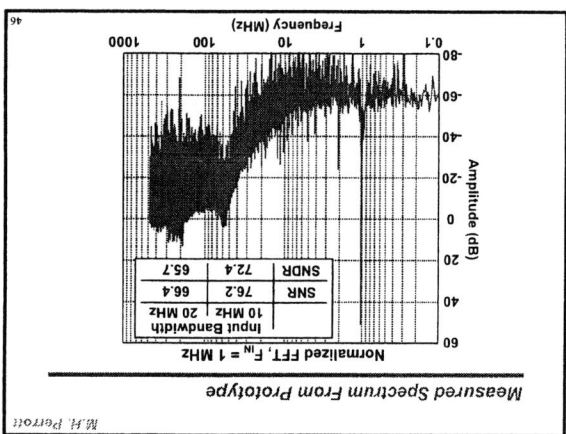

Measured Spectrum From Prototype

Normalized FFT, F_IN = 1 MHz

Input Bandwidth	10 MHz	20 MHz
SNR	76.2	66.4
SNDR	72.4	65.7

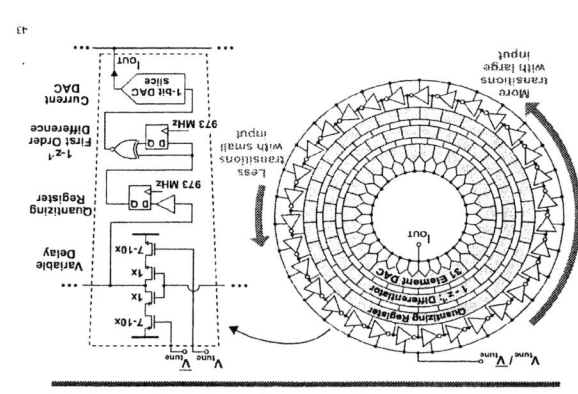

Measured SNR/SNDR Vs. Input Amplitude

SNR/SNDR vs. Amplitude, F_IN = 1 MHz

Conclusion

- Phase-locked loop circuits are about to go through dramatic changes as time-signaling is better utilized
  - New time-to-digital converter structures such as GRO
  - New structures/algorithms for achieving high performance (i.e., low noise, fast locking)
    - Example: MDLL circuit utilizing correlated double-sampling techniques to achieve low jitter
- It is time to apply this technology to other circuits, too
  - Example: ADC structures utilizing VCO-based quantizers

How can we further utilize time as a signal domain?

Low Power Reconfigurable Analog-to-Digital Converters

Kush Gulati

Cambridge Analog Technologies, Inc.
Bedford, Massachusetts

Sixth IEEE Dallas Circuits and Systems Workshop
November 2007

Outline

- Motivation and Applications
- Concept of Reconfigurable ADCs
- Prior Art for Reconfigurable ADCs
- Reconfigurable ADC Case Study
- Conclusions

Motivation for Reconfigurable ADCs

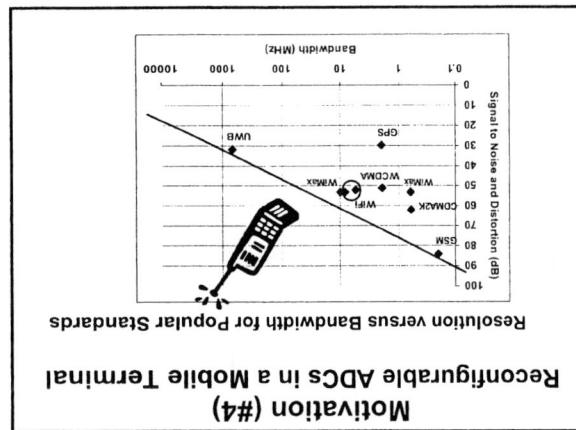

Resolution versus Bandwidth for Various Applications

Motivation (#2)

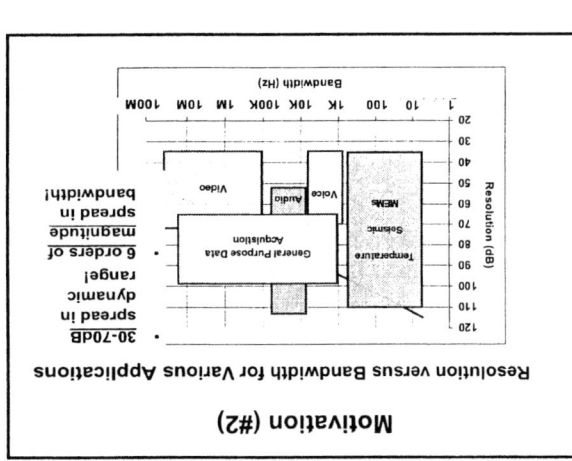

Resolution versus Bandwidth for Various Applications

- 30-70dB spread in dynamic range!
- 6 orders of magnitude spread in bandwidth!

Motivation (#3)

Multi-sensor battery operated wireless device
- Reconnaissance device
- Embedded medical applications

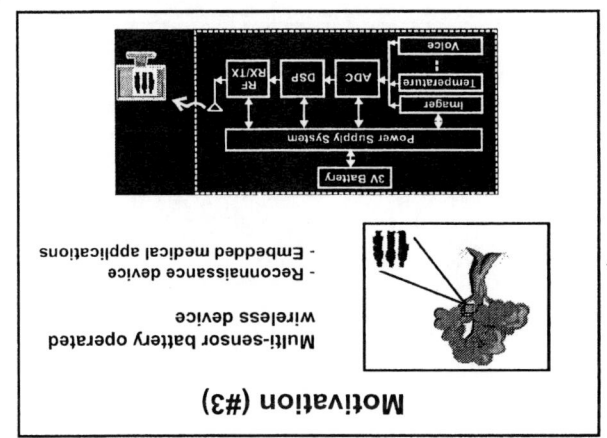

Motivation (#4)
Reconfigurable ADCs in a Mobile Terminal

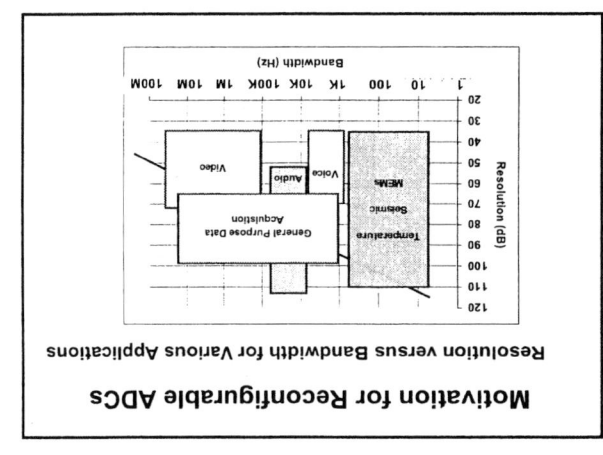

Resolution versus Bandwidth for Popular Standards

Motivation (#5): Reconfigurable ADCs in a Mobile Terminal

Resolution versus Bandwidth for Popular Standards

- 30-60dB spread in dynamic range!
- 2-4 orders of magnitude spread in bandwidth!

Motivation (#6): Solutions

- Multiple ADCs
  - Large area, high cost, not power optimized
- Single ADC with performance at highest common denominator
  - Technologically infeasible, high power consumption
- Reconfigurable ADC
  - Small area, power optimized

Motivation (#7): Power Versus Resolution

$$SNR \propto \frac{V_{SWING}}{\sqrt{\frac{kT}{C}}}, \; C=\text{Sampling Cap}$$

$$Power \propto C$$

$$Power \propto SNR^2$$

$$Power \propto 4^N, \; N=\text{Resolution (bits)}$$

In practice, $Power \propto 2^N \rightarrow 4^N$

Power consumption is exponentially related to ADC bits

Motivation (#8): Power Versus Bandwidth

$$G_M (\text{Transconductance}) \propto \frac{I_{BIAS}}{(V_{GS} - V_T)} \propto Power$$

$$G_M (\text{Transconductance}) \propto \frac{1}{\tau} \propto F_{SAMP} \propto 2 \cdot OSR \cdot F_{SIGBW}$$

$$Power \propto F_{SIGBW}$$

At constant OSR and device current density, power consumption is linearly proportional to input bandwidth

Motivation (#9)

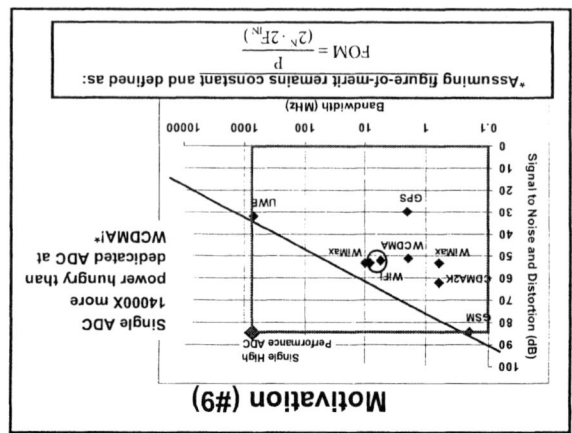

Single ADC 14000X more power hungry than dedicated ADC at WCDMA*

*Assuming figure-of-merit remains constant and defined as:

$$FOM = \frac{P}{(2^N \cdot 2F_{IN})}$$

Motivation (#10)

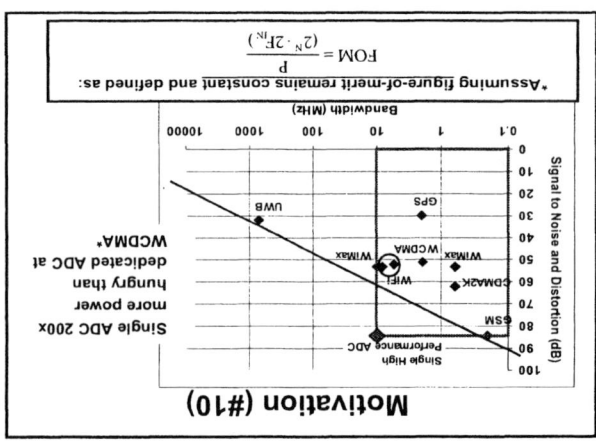

Single ADC 200x more power hungry than dedicated ADC at WCDMA*

*Assuming figure-of-merit remains constant and defined as:

$$FOM = \frac{P}{(2^N \cdot 2F_{IN})}$$

Solutions

- Multiple ADCs
  - Large area, high cost, not power optimized
- Single ADC with performance at highest common denominator
  - Technologically infeasible, high power consumption
- Reconfigurable ADC
  - Small area, power optimized

Advantages of Reconfigurable ADCs

- Reconfigurable ADC allows for optimized power consumption
- Reconfigurable ADC provides new opportunities, hitherto not possible
  - Variable quality of service
- Reconfigurable ADC enables faster time-to-market

Reconfigurable ADC Use Cases (cont.)

- Static reconfiguration
  - faster time-to-market
- Real-time reconfiguration (one at a time)
  - multiple standards/sensors
  - variable quality of service [1]
- Time-interleaved reconfiguration across different signals
  - simultaneous reception of standards/signals

Ref. [1]: S. Haykin, "Cognitive Radio: Brain-empowered wireless communications." IEEE J. Selected Areas in Communication, Feb. 2005

Outline

- Motivation and Applications
- Concept of Reconfigurable ADCs
- Prior Art for Reconfigurable ADCs
- Reconfigurable ADC Case Study
- Conclusions

Concept of Reconfigurable ADC (#1)

Resolution \ Data-Rate	Low <1K-100K	Medium 100K-10M	High 10M-1G	V. High >1G
High 15-20b	Delta-sigma	Delta-sigma		
Medium 10-15b	Delta-sigma	Delta-sigma Pipeline SAR	Pipeline	
Low 5-10b	Flash Algorithmic	Pipeline Flash SAR Algorithmic	Pipeline Flash	Flash

Vastly varying requirements of bandwidth and resolution best handled by different ADC architectures

- Architecture Reconfiguration
- Parameter Reconfiguration
- Bandwidth Reconfiguration

Concept of Reconfigurable ADC (#2)

Resolution \ Data-Rate	Low <1K-100K	Medium 100K-10M	High 10M-1G	V. High >1G
High 15-20b	Delta-sigma	Delta-sigma		
Medium 10-15b	Delta-sigma	Delta-sigma Pipeline SAR	Pipeline	
Low 5-10b	Flash Algorithmic	Pipeline Flash SAR Algorithmic	Pipeline Flash	Flash

ADC parameter tuning necessary within given architecture

- Architecture Reconfiguration
- Parameter Reconfiguration
- Bandwidth Reconfiguration

Outline

- Motivation and Applications
- Concept of Reconfigurable ADCs
- **Prior Art for Reconfigurable ADCs**
- Reconfigurable ADC Case Study
- Conclusions

Concept of Reconfigurable ADC (#3)

- Unity gain frequency of opamps tied to sampling rate of converter
- Opamp bias current needs to track clock frequency

- Architecture Reconfiguration
- Parameter Reconfiguration
- **Bandwidth Reconfiguration**

Reconfigurable ADC Prior Art

All Reconfigurable ADCs employ some combination of the following:

- Architecture Reconfiguration
- Parameter Reconfiguration
- Bandwidth Reconfiguration

Prior Art: Time-Interleaved Conversion [3]

Pros:
– Simplicity
– Opamp transistor current-density (& regime) is constant

Cons:
– Limited reconfigurability
– Large area

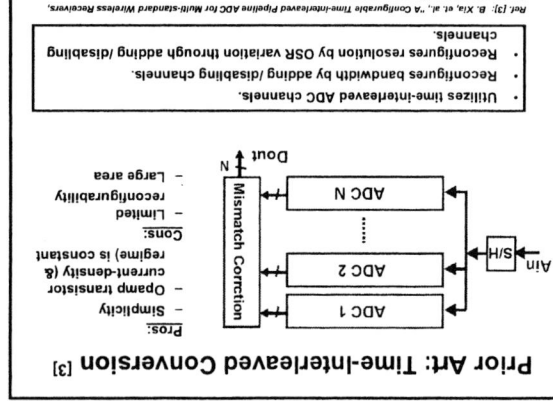

- Utilizes time-interleaved ADC channels.
- Reconfigures bandwidth by adding/disabling channels.
- Reconfigures resolution by OSR variation through adding/disabling channels.

Ref: [3]: B. Xia, et. al., "A Configurable Time-interleaved Pipeline ADC for Multi-standard Wireless Receivers," European Solid State Circuits Conference, Sep. 2004

Prior Art: Cascaded Delta-Sigma [4]

Pros:
– Small area
– Can change order without stability concerns

Cons:
– Cascaded ΣΔ has performance limitations
– ΣΔ architecture requires over-sampling of input

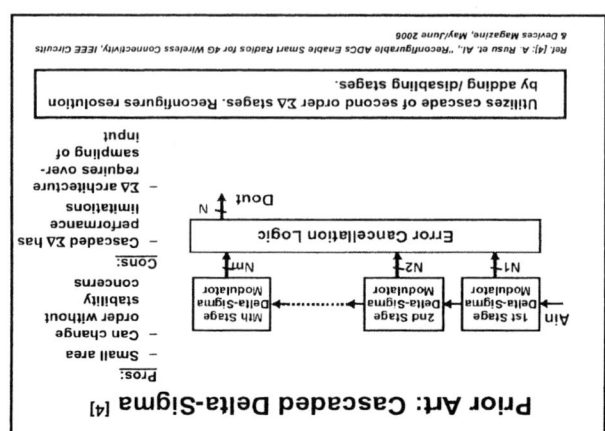

Utilizes cascade of second order ΣΔ stages. Reconfigures resolution by adding /disabling stages.

Ref. [4]: A. Rusu et. Al., "Reconfigurable ADCs Enable Smart Radios for 4G Wireless Connectivity, IEEE Circuits & Devices Magazine, May/June 2006

Prior Art: Pipeline/Cyclic Hybrid [5]

Pros:
– Wide conversion range

Cons:
– Because of stage scaling, cyclic converter can never be more power optimal than pipeline; cyclic mode is unnecessary [6]

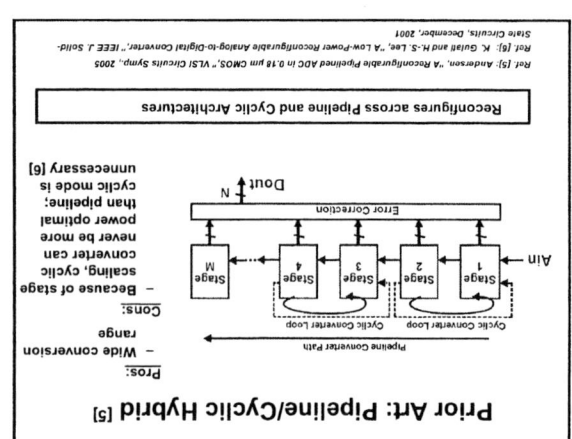

Reconfigures across Pipeline and Cyclic Architectures

Ref: [5]: Anderson, "A Reconfigurable Pipelined ADC in 0.18 µm CMOS," VLSI Circuits Symp., 2005
Ref: [6]: K. Gulati and H.-S. Lee, "A Low-Power Reconfigurable Analog-to-Digital Converter," IEEE J. Solid-State Circuits, December, 2001

Reconfiguration Methodology

- Architecture reconfiguration
  - Delta-sigma / Pipeline / Flash
- Parameter reconfiguration
  - Pipeline mode: capacitor size, pipeline length
  - Delta-sigma mode: OSR
- Bandwidth reconfiguration
  - Tune bandwidth of opamp automatically

Mapping Standard to ADC Mode (#2)

- Delta-sigma
  - GSM, EDGE, CDMA2K, Narrow-Band WiMax
- Pipeline
  - WCDMA, WiFi, Wide-Band WiMax
- Flash
  - UWB

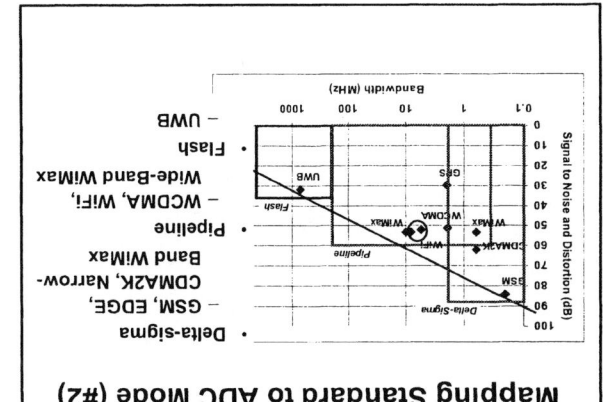

Mapping Signal to ADC Mode

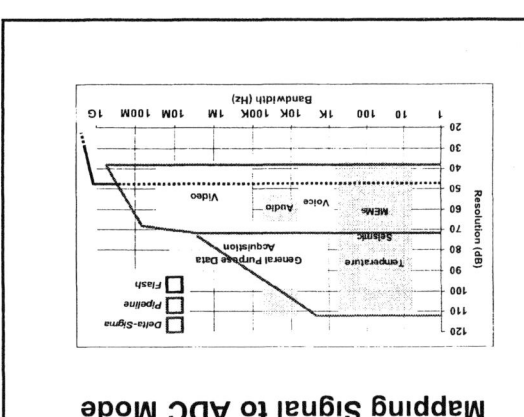

Reconfigurable ADC Case Study [6]: Concept

- Pipeline and Delta-sigma topologies work at different resolution / data-rate
  - Delta-sigma : low-medium bandwidth, medium-high resolution
  - Pipeline : medium-high bandwidth, low-medium resolution
  - Flash : very high bandwidth, low resolution
- Easy extension to Flash topology
- Combination of these topologies can cover wide resolution and data-rate
- These topologies share similar building blocks
- Build a number of common blocks and configure to desired topology

Ref. [6]: K. Gulati and H.-S. Lee, "A Low-Power Reconfigurable Analog-to-Digital Converter," IEEE J. Solid-State Circuits, December 2001

Outline

- Motivation and Applications
- Concept of Reconfigurable ADCs
- Prior Art for Reconfigurable ADCs
- Reconfigurable ADC Case Study
- Conclusions

Prior Art: Programmable Coefficient Delta-Sigma [7,8,9,10]

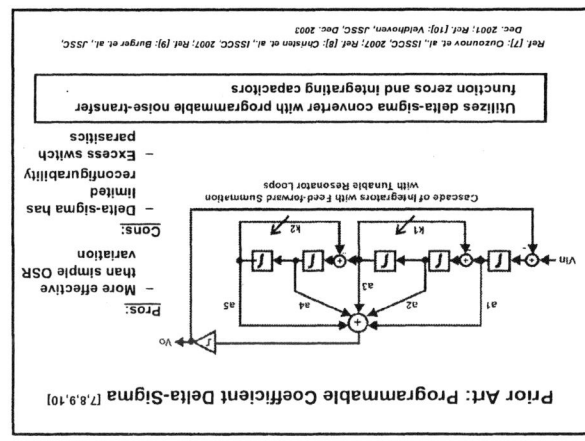

Cascade of Integrators with Feed-forward Summation with Tunable Resonator Loops

Pros:
- More effective than simple OSR variation

Cons:
- Delta-sigma has limited reconfigurability
- Excess switch parasitics

Utilizes delta-sigma converter with programmable noise-transfer function zeros and integrating capacitors

Ref. [7]: Ouzounov et al., ISSCC, 2007; Ref. [8]: Christen et al., ISSCC, 2007; Ref. [9]: Burger et al., JSSC, Dec. 2001; Ref. [10]: Veldhoven, JSSC, Dec. 2003

Overall ADC Architecture

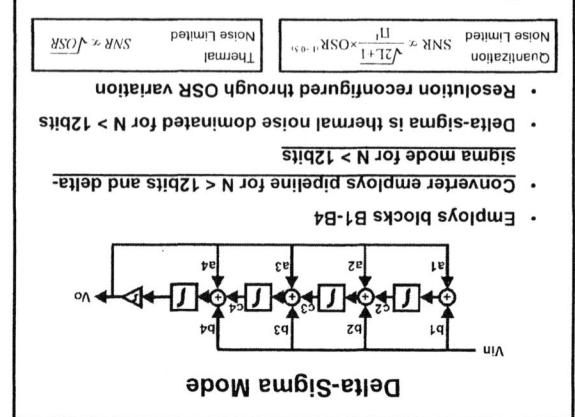

Config. Word — 5 · Fc · Clocks · PLL · Ain

Main Reconfiguration Logic — 8 · 3 · (Clocks) 12 · Block Logic · B1 · Block Logic · B2 · (Biases) 16 · Block Logic · BN

Interface Logic · Dout · Fckout — 16 · 4

Reconfigurable Block Architecture

- Composed of:
  - Opamp
  - Block switch fabric
  - Block capacitor array
  - Decision circuitry and logic
- Embodies:
  - 2 stages in pipeline mode
  - 1 stage in delta-sigma mode

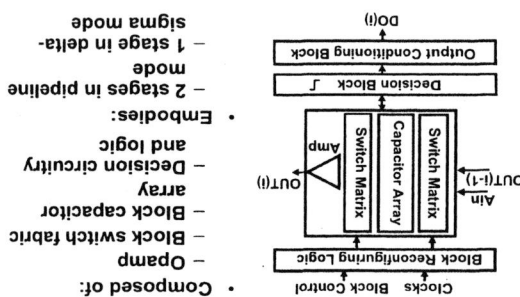

Clocks · Block Control · Block Reconfiguring Logic · Ain · OUT(i-1) · Switch Matrix · Capacitor Array · Amp · Switch Matrix · OUT(i) · Decision Block ∫ · Output Conditioning Block · DO(i)

Delta-Sigma Mode

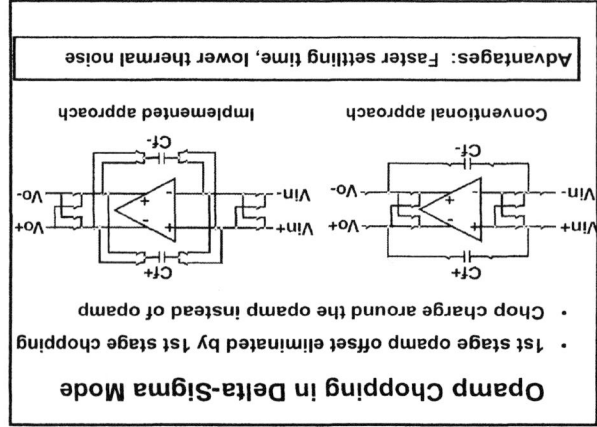

Vin · b1 ∫ a1 · b2 ∫ a2 · b3 ∫ a3 · b4 ∫ a4 · Vo

- Employs blocks B1-B4
- Converter employs pipeline for N < 12bits and delta-sigma mode for N > 12bits
- Delta-sigma is thermal noise dominated for N > 12bits
- Resolution reconfigured through OSR variation

Quantization Noise Limited	Thermal Noise Limited
$SNR \propto \dfrac{\sqrt{2L+1}}{\Pi^L} \times OSR^{L+0.5}$	$SNR \propto \sqrt{OSR}$

Opamp Chopping in Delta-Sigma Mode

- 1st stage opamp offset eliminated by chopping
- Chop charge around the opamp instead of opamp

Conventional approach   Implemented approach

Advantages: Faster settling time, lower thermal noise

Pipeline Mode

- Stage scaling of 2x between consecutive stages to minimize power
- Opamp shared between two consecutive stages to save power
  - 2 stages in 1 basic building block
  - inter-block scale factor = 4x
- Opamp offset and 1/f noise cancelled by Global ADC Chopping

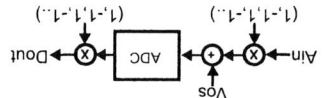

Ain ×(1,-1,1,-1...) + Vos → ADC → ×(1,-1,1,-1...) Dout

Pipeline Mode

- Capacitors can be reduced by 4x for 1 bit reduction
- Shift and truncate
- At different resolutions shift pipeline stages to maintain kT/C noise-limited operation for low power

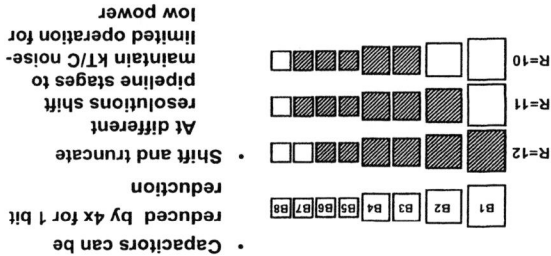

B1 B2 B3 B4 B5 B6 B7 B8

R=12
R=11
R=10

Flash Mode

- Multi-bit quantizer is typically placed at tail-end of pipeline [11] or multi-bit delta-sigma [12]
- Analog input bypasses basic building blocks to input of quantizer

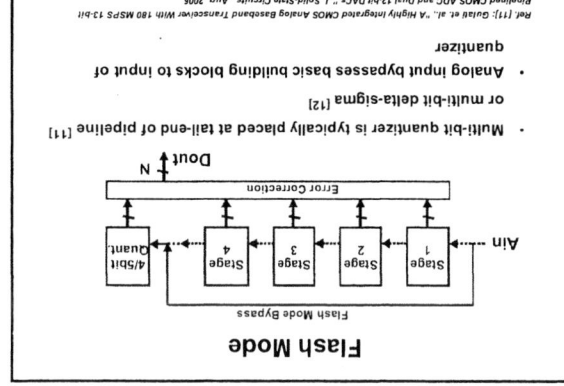

Ref. [11]: Gulati et. al., "A Highly Integrated CMOS Analog Baseband Transceiver With 180 MSPS 13-bit Pipelined CMOS ADC and Dual 12-bit DACs," J. Solid-State Circuits, Aug. 2006
Ref. [12]: Carley et. al., "Chapter 8: Delta-Sigma ADCs with Multibit Internal Converters," in Norsworthy et. al.,"Delta-Sigma Data Converters.", IEEE Press, 1997

Bandwidth Reconfiguration: Methods

- Look-up table (I_{BIAS} versus F_{clock})
  – Sensitive to process variation
- Opamp duty cycling [13]
  – Practical complexity
- Switched-cap method [14]
  – Inaccurate tracking across varying transistor regimes
- Change opamp device size with clock frequency
  – Inferior performance due to switch parasitics

Ref. [13]: Ahmed et. al., "A 50-MS/s (35 mW) to 1-kS/s (15 W) Power Scaleable 10-bit Pipelined ADC Using Rapid Power-On Opamps and Minimal Bias Current Variation," J. Solid-State Circuits, Dec. 2005
Ref. [14]: Andersen et. al., "A Cost-Efficient High-Speed 12-bit Pipeline ADC in 0.18-um Digital CMOS," J. Solid-State Circuits, July 2005

Bandwidth Reconfiguration Using PLL

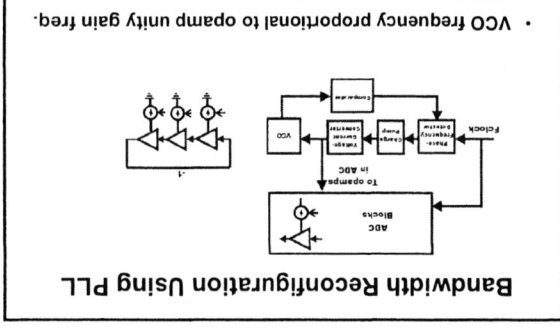

- VCO frequency proportional to opamp unity gain freq.
- VCO frequency tracks clock frequency
- Opamp UGF and settling time track clock frequency

Phase Locked Loop and VCO

- Cascaded scaled-down opamps as VCO
- Opamp bias current controlled by PLL

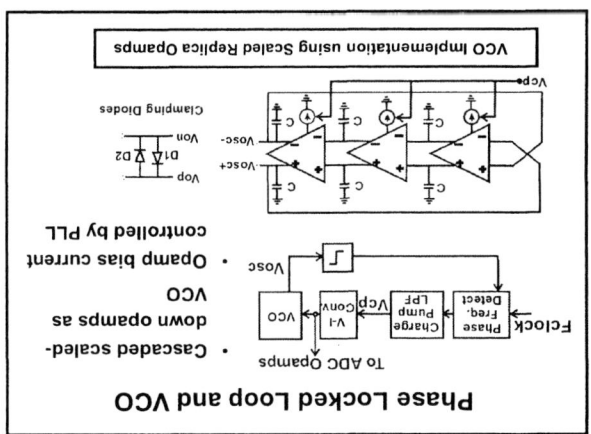

VCO Implementation using Scaled Replica Opamps

Typical Half-Block

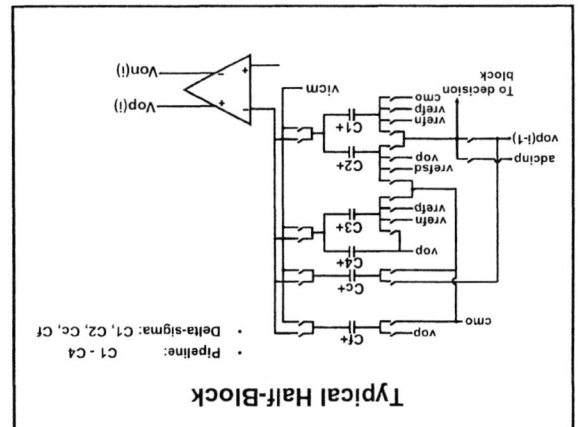

- Pipeline: C1 - C4
- Delta-sigma: C1, C2, Cc, Cf

Half-Block: Pipeline Mode

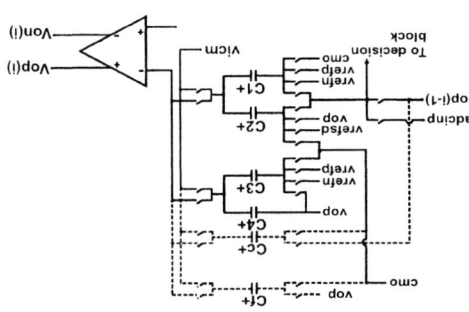

Conclusions

- Reconfigurable ADC critical to high performance multi-mode sensors / multi-mode radio
- Reconfigurable ADCs creates new opportunities
  - Additional power savings through variable quality of service
  - Superior time-to-market
- Truly reconfigurable ADCs require flexibility in architecture, parameter and bandwidth
- Sample case study demonstrates method for constructing flexible ADC
  - Employs pipeline / delta-sigma modes with possible extension to flash mode
  - Parameter tunability for SNR variation
  - Employs PLL for bandwidth reconfiguration

Comparison with Custom ADCs

$$FOM = \frac{2^{2N} \times (\text{Data Rate})}{\text{Power}}$$

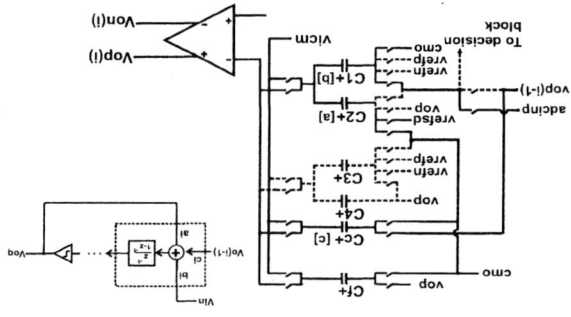

December 2001

PLL Measurements

Generated Ibias is exactly what is required for ADC opamps in all regimes of the input transistors

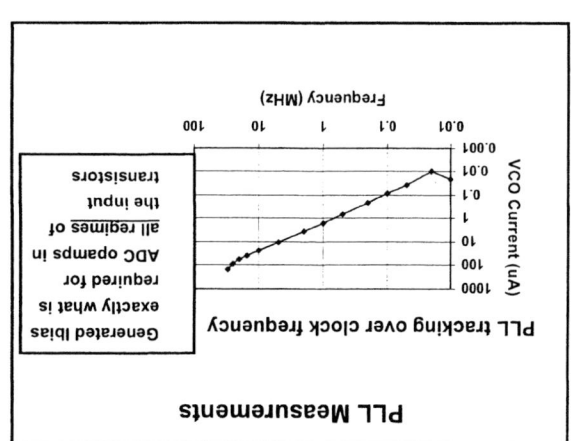

PLL tracking over clock frequency

Delta-Sigma Mode Measurements

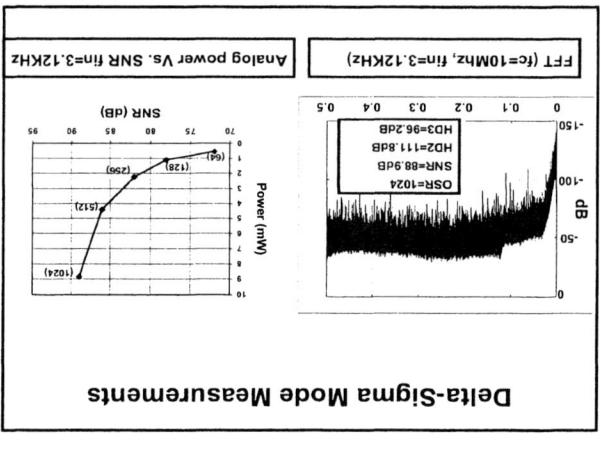

FFT (fc=10Mhz, fin=3.12KHz)

OSR=1024
SNR=88.9dB
HD2=111.8dB
HD3=96.2dB

Analog power Vs. SNR fin=3.12KHz

Generic Comparator

Threshold Requirements

Last stage (Pipeline & DS): zero

Intermediate stage (Pipeline): +/-Vref/4

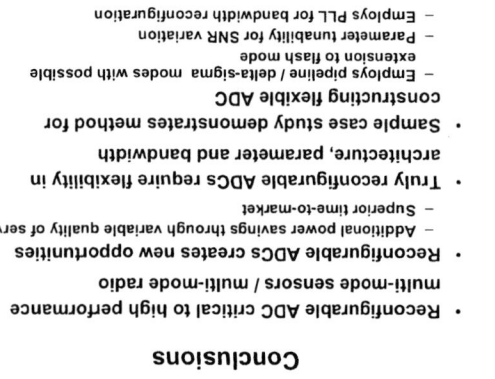

- Comparator can implement 0 or +/-Vref/4 thresholds
- Controlled by the block reconfiguring logic
- Sampling type comparator prevents kick-back noise

Half-Block: Delta-Sigma Mode

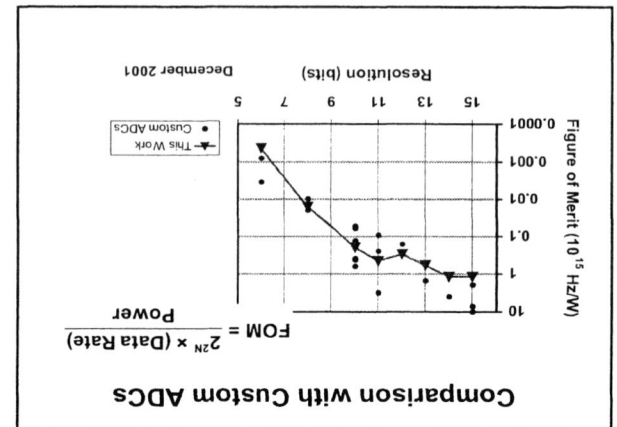

Designer's Guide Consulting
Analog, Mixed-Signal & RF Verification

Verification of Complex Analog Integrated Circuits

Ken Kundert
Henry Chang

Designs They Are A-Changin'
Bob Dylan

The Complexity of Design is Growing Rapidly

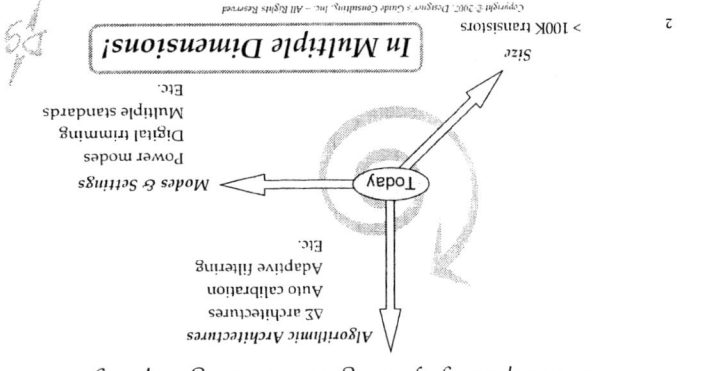

Size

Algorithmic Architectures
ΔΣ architectures
Auto calibration
Adaptive filtering
Etc.

Modes & Settings
Power modes
Digital trimming
Multiple standards
Etc.

Today

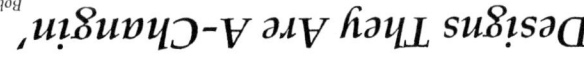

In Multiple Dimensions!

> 100K transistors

Copyright © 2007, Designer's Guide Consulting, Inc. – All Rights Reserved

2

Example: Audio Codec

> 10K transistors   > 30K cycles   > 250 settings

PGA ADC DAC PA Bias Control Interface

150 settings · 25 settings · 50 settings · 10 settings · 50 settings · 25 settings · 10 settings · Long time constant

Numbers given are representative of what would be common for this type of circuit.

Copyright © 2007, Designer's Guide Consulting, Inc. – All Rights Reserved

3

Why So Many Settings?

- Operating modes
- Power management modes
- Modes for specific customers
- Test modes
- Calibration
- Digital controls
- Digital trimming
  - These days, every resistor, capacitor, and current source is adjustable

Radj[0:7]   Cadj[0:7]   Iadj[0:7]

Copyright © 2007, Designer's Guide Consulting, Inc. – All Rights Reserved

4

Audio Codec Errors

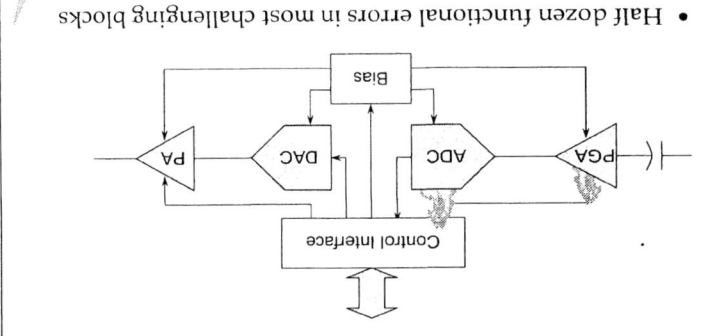

PGA ADC DAC PA Bias Control Interface

- Half dozen functional errors in most challenging blocks

Copyright © 2007, Designer's Guide Consulting, Inc. – All Rights Reserved

5

Functional Errors

In → Out

Enable

- Functional errors are often very simple errors
  - Inverted signals

Copyright © 2007, Designer's Guide Consulting, Inc. – All Rights Reserved

6

Functional Errors

- Functional errors are often very simple errors
  - Inverted signals
  - Corrupt logic

Functional Errors

- Functional errors are often very simple errors
  - Inverted signals
  - Corrupt Logic
  - Flipped busses

ctl[0:7] → ctl[0:7]

Functional Errors

- Functional errors are often very simple errors
  - Inverted signals
  - Corrupt logic
  - Flipped busses
  - Unaccounted for dependencies (chicken & egg problem)

Control Register → PLL, Bias

Functional Errors

- Functional errors are often very simple errors
  - Inverted signals
  - Corrupt logic
  - Flipped busses
  - Unaccounted for dependencies (chicken/egg problem)
  - Communication errors
- But are generally catastrophic

Design

The Two Basic Issues

- Detailed verification only performed at block level
  - All requires signals are assumed to be present
  - Assumptions on interblock dependencies never verified
- Verification on most modes never performed
  - Only typical or worst case modes
  - Any control logic that supports untested mode could contain hidden error

Transistor-Level Verification

- Too expensive for functional verification
  - 10K transistors, 30K cycles, 250 modes
  - One week for one mode with timing simulator
- Need nightly regression tests
  - 10K× speed up needed

Regression Testing

- Today, most designers test functionality at most once, when first designed
  - Redesign can break existing functionality
- In regression testing, we test all functionality every time tests are run
  - Greatly reduces risks with redesign

The Answer

- Exhaustive regression testing
  - Check every mode and every setting
  - Automated pass/fail tests (self-checking tests)
- Model-based verification
  - Dramatically accelerates the simulation
  - Moves it earlier in design cycle

Alternatives to Verification

Avoid functional errors by ...

- Limiting design choices, simplify circuits
  - Often not practical
  - Creates new risks
- Extensive use of conventions
  - Helpful, but not sufficient
- Exporting all control lines, fix bugs in firmware
  - Issues
    - Too many wires
    - Too much dependence of digital design group
  - Harder to bring up and test design
  - Not suitable for all issues (ex. chicken & egg problem)

Self-Checking Testbench

- Verification by visual inspection must be avoided
  - Errors can be very subtle
  - Too many tests, too many signals
  - Too time consuming, too error prone

Signal frequency is expected to change by 6%. Does it?

Exhaustive Testing Example

```
task checkRXgain;
  real linGain, prevGain; integer seed;
  begin
    $display("gain set to %d", gain);
    linGain = pow(10, (gain - 20)/20);
    out = $dist_uniform(seed, -1M, 1M)/2M;
    in = out / linGain;
    #(settlingTime);
    measGain = V(pout, nout) / in;
    $display("Gain is %R", measGain);
    if (abs(V(pout,nout) - (out)) > 50m)
      $display("FAIL: gain out of range");
    else if (measGain < prevGain)
      $display("FAIL: gain not monotonic");
    else $display("pass: gain okay");
    prevGain = measGain;
  endtask
```

```
initializeTestbench();
initializeDUT0;
pnp = 0;
#(2*clockPeriod);
checkSupplyCurrent0();
pnp = 1;
#(200*clockPeriod);
checkSupplyCurrent0();
for (i = 0; i < 64; i=i+1) begin
  gain = i;
  checkRXgain0();
end
...
```

Check Power Modes

Check Gain Settings

What's Needed

- Systematic approach to verifying design & specification
- Confidence that all flaws have been found
- More verification, earlier in design flow
  - Errors are easier to fix & less disruptive
- Help with performance verification
- Accurate model of MS section

Model-Based Verification

- A systematic approach built on two important concepts
  - Mixed-signal simulation
    - Allows efficient co-simulation of analog and digital sections
    - Dramatically accelerates simulation
  - Mixed-level simulation
    - Dramatically accelerates simulation

Mixed-Signal Simulation

- Combined logic and circuit simulation
- Based on Verilog-AMS or VHDL-AMS
- Verilog-AMS
  - Combines Verilog and Verilog-A, plus a bit more
    - SPICE
- VHDL-AMS
  - Adds analog extensions to VHDL (there is no VHDL-A)

Verilog-AMS Example

```
module flash( out, in, clk, bias, pwrdn, vdd );
    input in, clk, bias, pwrdn, vdd;
    output [15:0] out;
    electrical in, bias, vdd;
```
• Can have both logic and electrical ports, adds real-valued event-driven ports

```
    ...
    initial begin       // Verilog
    ...
    end
    always begin        // Verilog
    ...
    end
    analog begin        // Verilog-A
    ...
    end
endmodule
```

- Initial/Always (Verilog) sections run in the event driven (digital) kernel
  - Analog signals, variables, and events can be observed
- Analog sections run in the SPICE kernel
  - Digital signals, variables, and events can be observed

"Connect Modules" (special modules) are automatically inserted as needed to convert digital to analog signals and vice versa

---

Mixed-Level Simulation

- Replace transistor-level circuit with model
  - Dramatically accelerates simulation
  - Verification can start before schematics are available
  - Model can be used for system level verification, test development, etc.

But how does one assure model matches implementation?

Model Verification

Apply the same tests to both

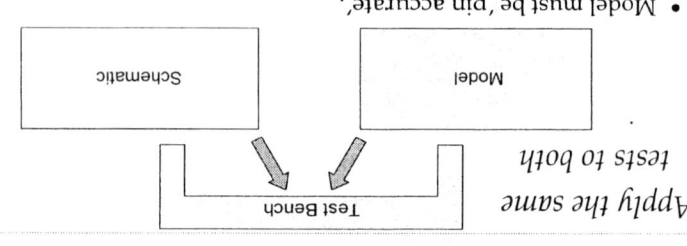

- Model must be 'pin accurate'.
- Model can be developed before schematic
  - Schematic designed to match model
  - Model becomes the 'specification' or target
  - Naturally encourages top-down design
- Generally takes too long to simulate with full schematic

Accelerating Model Verification

Decompose model into blocks.
Then perform mixed-level simulation.

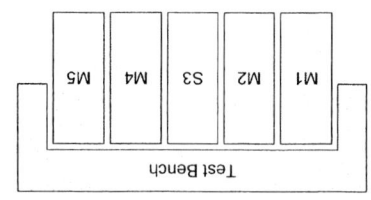

- Replace one block at a time with schematic

Still Too Slow?

Repeat the process recursively.

- Each model must be pin-accurate
- Hierarchically decomposes verification

Example Simulation Times

- Mixed-level sim time for loopback test

Representation	Tests	Time
Verilog	250	2 min
Verilog-AMS	250	6 min
PGA @ xstr	150	6 hour
ADC @ xstr	50	12 hours
Bias @ xstr	25	1 hour
DAC @ xstr	10	6 hours
PA @ xstr	50	3 hours

- Tests can be run simultaneously
  - Three computers enough to verify overnight

Numbers given are representative of what would be common for this type of circuit.

Controlling Simulation Time

- To reduce duration of regression tests
  1. Further partition blocks so as to require fewer transistors in a mixed-level sim.
     - Cost is increased model/TB development
     - Cost is more licenses, computers
  2. Partition tests
     - Cost is more licenses, computers
- Run in parallel on multiple computers
- Goal is to complete all tests by morning

Writing Models

- Functional models are generally simple and easy to write
- Model requirements
  - Pin accurate - allows mixed-level simulation
  - Models behavior of all bias and control signals
  - Models function
  - Can add more for performance if needed
- Add assertions for more checking

Assertions

An assertion is a condition monitored passively that must always be true

- Useful for monitoring …
  - Setup & hold conditions
  - Bias conditions
  - Transfer functions
  - Illegal states
  - Most design assumptions
- Parallels use in digital verification
- Catches errors in often ignored signals

Modeling Pins (with Assertions)

What is modeled	Explanation
Supplies	Validate the supply is in the expected range, Model expected power consumption in all modes
Clock inputs	Is it the right frequency? Check setup & hold violations
Bias input	Thevenin equivalent; check that bias is within tolerance
Bias output	Thevenin equivalent; check that it operates within range
Analog inputs	Verify that the signal input is within compliance range
Analog outputs	Model output drive capabilities as needed
Digital signals	Check for x or z
Signals for test	Functionality modeled

Example: Flash ADC

```
module flash_adc ( out, in, clk, bias, pwrdn, vdd );
input in, clk, bias, pwrdn, vdd;
output [15:0] out;
electrical in, bias, vdd;
integer i, level;
reg pwrf, biasFault;
reg [15:0] d;
always @(posedge clk) begin
    pwrFault = (V(vdd) > 1.9) || (V(vdd) < 1.7);
    biasFault = (I(vdd,bias) > 16u) || (I(vdd,bias) > 14u);
    level = 16*(V(in)+0.5);      // convert input to an integer
    for (i=0; i<16; i=i+1)
        d[i] = (i < level);
end
assign out = (pwrdn || pwrFault || biasFault) ? 16'bx : d;
analog begin
    V(vdd,bias) <+ pwrdn ? 0 : 0.5 + 20k*I(vdd,bias);
    I(vdd) <+ pwrdn ? 1u : 500u;
end
endmodule
```

- I/O declarations
- Internal variables
- Assertions for power and bias
- Functional model
- Generate the output
- Model the bias input
- Model power consumption

Verification Methodology

- Develop verification plan
  - To verify specs
  - To verify concerns (brainstorm)
- Develop modeling and simulation plans
  - Identify which models need developing
    - May be more than one for each block
- Create top-level schematic and models
- Develop autonomous (self checking) testbench
- Perform mixed-level simulation
- Automate regression testing

Outcome

- Verified design
- Verified spec – important for large teams
- Verified top-level model – critical for IP
- Control of risks enables greater innovation
- Reduced need for senior designers
- Encourages top-down design
- Aids reuse, test development

Top-Level Model

- "Sign-off" quality required
  - Digital verification engineers used to Verilog models
  - Digital methodology requires trustworthy models
  - Must match the functionality (all modes) exactly
  - Delivering bad model worse than delivering no model at all
- Generally must be Verilog or VHDL
  - Digital designers unwilling to take AMS models
- Needs an adaptor to fit into top-level testbench
  - Top-level testbench written for AMS models

This is Analog Verification

- Exhaustive regression testing
- Traceable to transistor level
- Verifies both models and circuits
  - Test benches verify behavior of models
  - Mixed-level simulation assures models are consistent with circuit
- Driven by verification engineer

Analog Verification Engineers

- Why?
  - Separation of design and verification
  - Verification needs focus designers cannot give
  - Verification needs skills designers often do not have
    - Modeling & scripting
    - VEs also need design skills; could be designers in training
- Role
  - Capture verification plan
  - Write models
  - Write, debug, & run test benches
  - Track bugs

Verification Plan Example

- Must always be satisfied (assertions)
  - Power consumption okay
  - Common-mode output voltage okay
  - Correct bias voltages & currents at all blocks
  - Correct supply voltages at all blocks
  - No setup/hold violations on control registers
  - Etc.
- Must be satisfied in specific situations (tests)
  - Power consumption in power-down mode
  - No pop upon power-up
  - Gain monotonic and within tolerances
  - Etc.

Writing the Model & Testbench

- Read spec, interview designer
  - Determine basic function of each block & description of each pin
  - May need to supplement, or actually write, specification
- Use verification plan to write model
  - Give careful attention to needed assertions
  - Do not over model
- Let's step through the process ...

Writing a Model Example

- Step 1: Capture basic behavior

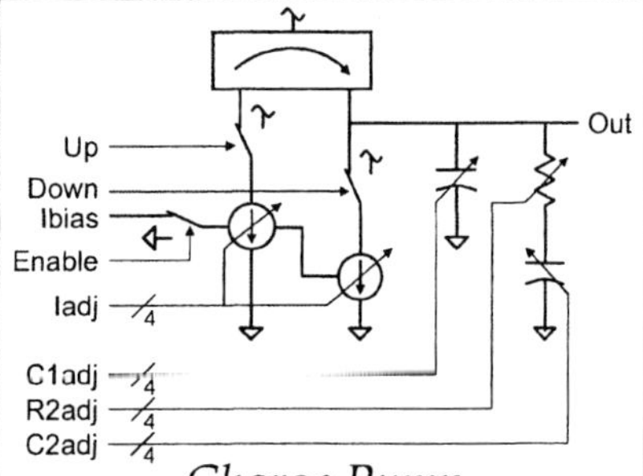

Charge Pump

Writing a Model Example

- Step 2: Capture pin descriptions

Pin	Dir	Type	Description
UP, UPb	in	Pseudo-differential	Sources Icp to output
DN, DNb	in	Pseudo-differential	Sinks Icp from output
Ibias	in	10 µA sink, ± 5%	0.5 V @ 10 kΩ nom.
enable	in	Boolean, active high	Activates CP
Iadj	in	4-bit binary	Sets Icp: 0 - 500 µA
C1adj	in	4-bit binary	Sets C1: 0 - 100 pF
R2adj	in	4-bit binary	Sets R2: 0 - 100 kΩ
C2adj	in	4-bit binary	Sets C2: 0 - 100 pF
out	out	Analog voltage	Rail to rail
Vdd, Gnd	pwr	1.8 V dc supply, ± 10%	1.2 mA max.

Writing a Model Example

- Step 3: Identify assertions
  - $Vdd = 1.8\ V \pm 10\%$
  - $Ibias = 10\ mA \pm 5\%$ when enabled
- Step 4: Write the model
  - Excerpts shown next

Writing a Model Example

```
analog begin
  // charge pump
  state = (enable && !fault);
  I(out) <+ Ion*transition(state, 0, tr);
  // filter
  I(out) <+ c1 * ddt(V(out));
  V(out,int) <+ r2 * I(out,int);
  I(int) <+ c2 * ddt(V(int));
  // supply
  I(vdd) <+ enable ? 1.2m : 0;
  // bias input (norton equivalent)
  I(Ibias) <+ enable ? (V(Ibias) - 0.4)/10k : 0;
end
```

```
always @(UP or DN or model charge pump) begin
  if (UP && !DN) state = 1;
  else if (!UP && DN) state = -1;
  else state = 0;
end
always begin // set output current value
  Ion = 31.25u * Iadj;
  @(Iadj);
end
always begin // assertions
  vddFault = (V(vdd) > vddmax) ||
             (V(vdd) > vddmin);
  IbiasFault = (V(Ibias) > Ibiasmax) ||
               (V(Ibias) < Ibiasmin);
  @(UP or DN);
end
assign fault = vddFault || IbiasFault;
... etc
```

Writing a Model Example

- Step 5: Write test bench for model
  - Not shown (similar to earlier example)
- Step 6: Perform mixed-level simulations
  - Adjust testbench tolerances to fit circuit
  - Resolve any remaining differences
    • If circuit correct: fix spec, model, testbench
    • If testbench correct: fix circuit
- Repeat for all models
- Repeat for top-level Verilog model
  - Top-level testbench must satisfy verif. plan

Analog Verification Engineers

- Lead Verification Engineer
  - Peer to design lead
  - Has ultimate responsibility for functional silicon
  - Verification, simulation, & modeling plans
  - Track bugs, track changes to spec, help manage TLS, manage models, test benches
- Individual Verification Engineers
  - Writes models, test benches
- Typically 1 VE needed for every 5 DE
  - Depends strongly on number of modes in design

Adoption

- Phase 1: Verification separate from design
  - Designer's not dependent on models, testbenches created by verification engineers
  - Goal: build understanding, experience & trust; minimize impact & risk on design
- Phase 2: Integrate verification & design
  - Designers dependent on models, testbenches
  - Natural with derivative designs
  - Goal: maximize efficiency of design team
  - Leads to top-down design

Integration into Top-Down Design

- Verification does not require TDD flow
  - Layer over existing flows with little disturbance
- Forms symbiotic relationship with TDD
  - Verification methodology provides models that aid TDD
  - Models help to close gap between system design and circuit design
  - TDD provide top-level schematic and clean interfaces that smooth verification process

---

Review

- Analog circuits have changed
  - Large complex & 1000s of modes common
- Methodical verification of these circuits is possible
  - And will soon be expected
- Model-based verification is only approach
  - Everything else is *much* too slow
  - Based on mixed-level simulation
- Benefits from verification engineers
- Naturally complements TDD
- Substantial change, tough to get started
  - Call me if you need help

References

- H. Chang & K. Kundert, "Verification of Complex Analog and RF IC Designs." *The Proceedings of the IEEE*. To be published in 2007.
- K. Kundert & H. Chang, "Verification of Complex Analog Integrated Circuits." CICC-06.
- K. Kundert & O. Zinke. *The Designer's Guide to Verilog-AMS*. 2004.
- K. Kundert. "Principles of top-down mixed-signal design." *www.designers-guide.org/Design*.
- A. Meyer. *Principles of Functional Verification*. 2003.

TOWARDS EFFICIENT SPECTRUM SHARING

Borivoje Nikolić
Berkeley Wireless Research Center
University of California,
Berkeley, CA 94720-1770

Abstract—One hundred years of spectrum sharing based on fixed frequency allocations have led to fracturing and poor utilization. Present methods of frequency allocation combined with a reliance on fixed infrastructure threaten to halt this growth. An additional consequence is the deployment of fundamentally less robust systems, prone to disruption in major disasters or overload. This paper outlines a vision of a system that, by enabling the secondary use of spectrum on an opportunistic basis and by establishing collaboration between the users, would enable a realization of ubiquitous, robust and agile wireless systems that are able to support further traffic growth and changing demands in traffic.

Abstract—**Wireless communications, radio, transceivers, CMOS.**

I. INTRODUCTION

Wireless communications are experiencing an explosive growth in the 21st century. Reliance on the mobile telephone for daily voice and data communication, and often for first contact in case of emergency, has become pervasive. Mobile telephones are widely deployed and are as commonly carried in our pockets or purses as wallets and house keys. Simultaneously, wireless LAN is presently included with all notebook computers and many portable computing, communications, and entertainment devices. Future wireless communications outline a growth path for better and more diverse coverage, higher datarates, and increased functionality over existing cellular, local area and personal area networks. All the new and growing applications require more bandwidth, and scarcity of the available spectrum jeopardizes this vision. On the other hand, wireless networks of today are prone to disruption, particularly in cases of sudden surges in usage or major disasters.

This paper outlines a vision of a system that: (i) enables the secondary use of spectrum on an opportunistic basis and (ii) establishes collaboration between the terminals. The terminals in this system will operate in a very broad frequency spectrum with bands of operation that can be dynamically allocated. Such a system would be able to reuse the frequency bands that the primary users are not using at a particular time and a particular location. The terminal will be able to migrate from infrastructure-supported operation to communication within a mesh network, using either centralized frequency allocation or intelligent and cooperative sensing of unutilized bands. The concept of secondary use of the spectrum, known as cognitive radio [1][2], in combination with advanced cooperation between system components is enabled by advances in fundamental communications and networking theory and continued improvements in integrated circuit technology. Because of its universal functionality, the same device can be used for many applications that otherwise require dedicated and expensive low-volume solutions. It is envisioned that a system like this would be able to support several modes of use:

- Commercial, high datarate, multi-mode communications.
- Public safety networks
- Communications in case of major disasters in true cognitive manner.

Commercial cellular networks benefit from large volumes of devices manufactured every year to bring the cost per user down. On the other hand, public safety radio communications are locked into separate frequency bands, which limit them to the use of low-volume and expensive solutions. The envisioned system would provide first responders with the large bandwidth necessary to transmit high date rate signals (such as video) from a large number of sources. Finally, in a case of a major disaster, where millions of people may get affected, with communication infrastructure disrupted, a system like this would provide expanding capacity needed for emergency communications.

In this paper, a characteristic features that limit the growth of wireless communications today are discussed first, followed by a discussion about the cognitive collaborative wireless systems. The main technological limitations that need to be overcome are discussed next, and the paper is concluded with a brief summary.

II. CHARACTERISTICS OF WIRELESS COMMUNICATIONS TODAY

A striking characteristic of the 'connected' society is that a huge majority of the population now owns multi-standard, long-range, potentially location-aware radios in the form of mobile telephones. This device can have many uses beyond daily commercial communications; of particular interest are the uses by the public safety and by general population in a major disaster. The majority of people in need at a time of the disaster will have these devices with them; but they will not be able to utilize them, as the infrastructure they rely on to operate will not be useable.

Growth of commercial wireless communications and robustness of emergency services trace their challenges to the same set of obstacles:

(i) Rigid principles of spectrum sharing;

(ii) Traditional narrowband nature of radio transceivers, dictated by the underlying technology;

(iii) Limited cooperation between different or similar devices, systems and standards;

(iv) High dependence on wired infrastructure to operate wireless systems.

i) Rigid spectrum sharing. Spectrum is, by its very nature, a shared resource. The traditional approach of spectrum sharing is based on explicit spectrum division and licensing for a particular use. The use of spectrum as a shared resource is regulated by government agencies in each country (Federal Communications Commission, FCC, in the U.S.). Frequencies are assigned on a permanent basis and are very rarely re-allocated for different use. Within a licensed band, the spectrum is either explicitly divided among uses and users, or a restriction is imposed on the power transmitted by any individual user within a particular band. While these approaches have the advantage of conceptual simplicity and ease of enforcement, they do not scale with the demand for growing and evolving wireless communications.

One hundred years of spectrum sharing based on fixed frequency allocations have led to fracturing and poor utilization. Even when licensed, the spectrum is often not used; particularly when observed in a particular space and at a specific time instant. Allocating a particular frequency band for the exclusive use of a single service wastes that frequency band whenever and wherever there happens to be insufficient demand for that service. Moreover, the use of the spectrum becomes inefficient whenever the technology delivering that service becomes antiquated.

The present way of spectrum sharing hampers the growth of wireless communications, because it s not scalable in terms of the number of simultaneous users and datarates:

• To enable higher datarates in wide-area networks operating with relatively narrow frequency bands, it is necessary to increase the number of base stations; which is not possible in most U.S. urban areas.

• It locks certain applications, like public safety, into particular bands where, because of the low volume of devices being manufactured, they cannot benefit from rapid technology advances available to high volume applications like mobile telephony.

• Centralized frequency control leads to infrastructure-based systems, which are fundamentally vulnerable to disruptions, particularly in the case of disasters and malicious attacks.

• The number of devices that can simultaneously operate is limited, which presents a major limitation in cases where many people need to communicate.

In Figure 1, a snapshot over 50μs at a sample rate of 20GS/s in downtown Berkeley, shows that the TV, FM radio, cellular and unlicensed bands are well utilized, while the spectrum usage at frequencies from 3-6 GHz is minimal [3]. The existing approach of allocation largely ignores two other degrees of freedom for sharing besides the frequency, the space and the time, and results in a spectrum scarcity in the presence of this spectrum abundance.

ii) Narrowband radio technology. All radio receivers since Armstrong's can be viewed as methods of narrowband filtering that precedes a detector (or an analog-to-digital converter). This architecture matches well the traditional radio system specification of narrowband operation in a fixed frequency band. However, proliferation of modes and standards in modern communications has driven the demand for changes in radio architectures that can support multi-mode communications. Modern communications terminals need to implement WAN, LAN, PAN operation together

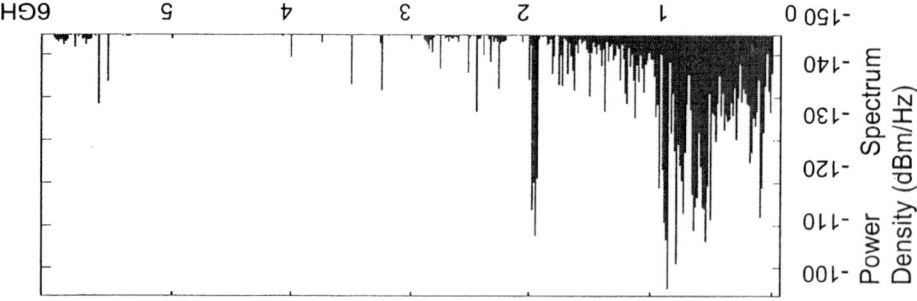

Figure 1: A typical radio spectrum measured over 50μs using one-million-point FFT in downtown Berkeley [3].

with positioning, while often supporting more than 20 different standards.

iii) Limited collaboration in communication systems. Rigid spectrum sharing, together with narrowband communications technology has contributed to the development of wireless systems with very little collaboration. Cooperation between the terminals within one system is often limited to establishing only peer-to-peer links, following a centralized allocation of resources by the access point or base station. Furthermore, cooperation between different wireless systems is often limited to coexistence in the same frequency band. This system concept greatly limits its capacity.

iv) Dependence on wired infrastructure. Wireless communication, by the nature of the signal transmission, promises high robustness to disruptions. However, this inherent robustness is limited by today's common practice of needing the wired infrastructure to achieve connectivity between wireless components. The heterogeneity of wireless systems in operation today spans a wide range of cell sizes, from peer-to-peer communications over local-area networks, to wide-area networks. Since they are all operated under different standards, there is no collaboration between these systems except the attempt to avoid causing each other interference.

III. COGNITIVE AND COLLABORATIVE RADIOS

A dramatic shift in wireless systems design occurs with the introduction of concepts of cognitive radios with collaboration between the terminals. The concept of a cognitive radio generally covers the secondary use of a spectrum allocated to a primary user on an opportunistic basis. In contrast to the ultra-wideband (UWB) transceiver, that presents the spectrum underlay by operating below the noise level, cognitive radio overlays the spectrum by operating in the "white space" between the primary users. A spectrum of scenarios for operating cognitive radios can be envisioned, from the spectrum lease by primary users on a temporary basis, to unlicensed use of unused spectrum based on sensing the primary users. In many of these scenarios, it is envisioned that primary users, such as public safety systems will have the "right of way" and can deploy "lights and sirens" to request that non-critical systems yield spectrum.

The complementary concept of collaboration between the terminals within a cognitive system can dramatically improve the wireless system efficiency. Terminals in today's systems are generally unaware of each other; they communicate with the base station/access point in a centralized manner, by using allocated frequency channels and time slots. In many cases, neighboring terminals present worst-case RF interferers to each other. Concept of collaboration encompasses a broad spectrum of ideas. Terminals aware of each other's proximity can use that information to minimize interference through cooperation; in an extreme case, collaborating terminals can expand the system capacity by using packet-multihop or collaborative diversity.

IV. COGNITIVE RADIO TECHNOLOGY

Implementation of a cognitive radio with collaboration requires changes in all layers of wireless system, from physical layer up to the policy.

Spectrum sharing. One remarkable feature of the recent growth in wireless usage is that many of the new devices today are operating in the "unlicensed" parts of the spectrum that are explicitly meant to be shared for many uses. This openness to new users and uses spurred the deployment of new wireless systems. However, the amount of unlicensed spectrum currently used is limited and the way it is being shared it is far from perfect [4]. Co-existence of various systems that share the same band has not been cared of by their design, and has to be resolved in practice [5][6][7]. Because of legacy spectrum allocations there is little opportunity for opening new spectrum bands for licensed or unlicensed use. The most notable exception is 7GHz of bandwidth that has been opened for unlicensed use worldwide in the 60GHz band. However, propagation characteristics of the mm-wave frequencies favor indoor operation.

Judged by their recent moves and studies, the FCC appears to be receptive to secondary use of the spectrum; however, this use has been cautiously controlled. In a move to reclaim poorly utilized spectrum, the FCC has started allowing secondary use of frequency bands already allocated to primary users. UWB systems are allowed to operate in the 3-10GHz range given that their power spectral density is below the level of noise [8]. The FCC has also allowed for the secondary use of the unused channels in the UHF TV band, 400-800MHz [9], and a dual use of 700MHz band for commercial and public safety applications [10].

Radio architectures. The envisioned cognitive radio system necessitates the use of a wideband, frequency-agile wireless transceiver. Advances in the underlying communications and computing technology have evolved to enable us to devise dramatically different radio architectures from the one that has been in use for past 80 years. In order to operate in a wider spectrum, a new, broadband radio architecture is needed. CMOS has been the technology of choice for commercial wireless systems for the past decade. Scaling of the CMOS technology increases its operating frequencies and reduces the device capacitances. This enables the design of radios that operate at higher frequencies and wider bandwidths, but with decreased linearity. Deeply-scaled CMOS, augmented by other technologies, such as MEMS can provide alternate means of filtering RF signals [15]. Such a wideband receiver is very sensitive to in-band interference, and active interference cancellation has to be embedded into the architecture [11][12].

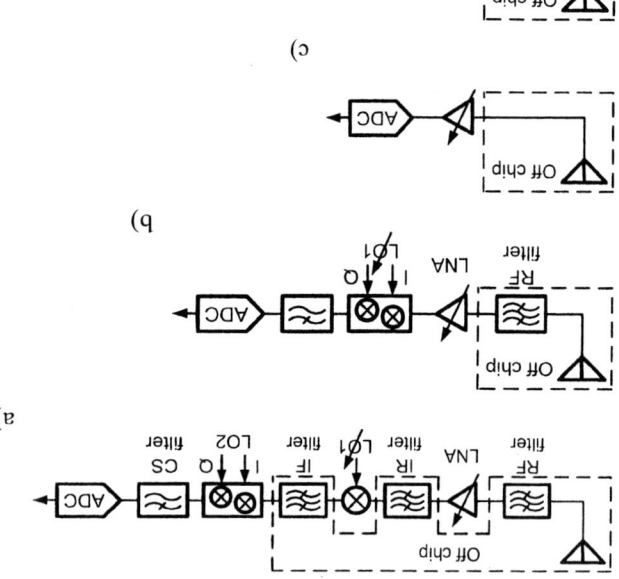

Figure 2: Receiver architectures: a) super-heterodyne, b) direct conversion, c) software-defined, d) wideband.

The wideband radio architecture will evolve with technology; over time, it would be able to communicate over a broader band, and be more agile as technology advances.

In addition to the wideband and frequency-agile front-end, the wireless transceiver will need to have a dynamically configurable, wideband, power-efficient baseband processor.

Wireless receiver architectures have migrated from super-heterodyne with high IF frequency, Figure 2.a, to direct conversion with zero or low IF, Figure 2.b, to minimize the number of off-chip components in CMOS realization. Migration to wideband operation will require a further evolution of receiver architectures. Wideband operation is necessary in order for the radio to be able to scan over a wide spectrum and utilize unused spectrum in an opportunistic fashion. For instance, one can envision a transceiver operating in the 1-10 GHz band and continuously monitoring unlicensed WiFi, WiMax and UWB bands to maximize throughput and minimize power consumption. This requirement is difficult due to the widespread use of tuned elements (such as inductors) in the design of present day transceivers. On the other hand, a common vision of a software-defined radio [1], Figure 2.c presents a major challenge for the ADC design. Removal of the narrowband filtering before the ADC dramatically increases the dynamic range of the signal, which, when coupled with the wide frequency band results in a very large power requirement. Technology scaling has been simultaneously improving the sampling frequency and power of ADCs; but the rate of progress removes this architecture from consideration in near future. Instead, a wideband architecture can be envisioned in near future, where the tuned filters will be replaced with inductorless wideband filters, aided with active interference cancellation [13][14], to reduce the dynamic range requirements for the ADC, Figure 2.d.

Collaborative communication. Sixty years of information theory have led to understanding the capacity of individual point-to-point channels, its relationship to physical media, and methods to achieve channel capacity in practical systems. However, the idea of overall system capacity, a notion that is more relevant to shared media is largely underdeveloped. By using cooperation between the devices it is possible to greatly increase system capacity. Collaboration between wireless nodes can be used to reduce the probability of causing harmful interference in truly cognitive modes of operation. A single device has an unacceptably large chance of missing a primary user of the frequency band because of the finite chance that the received signal strength will be below the receiver's sensitivity, due to local fading. Multiple terminals, in collaboration with a communications broker have a better chance of detecting primary users.

Figure 3 describes how the aggregate capacity of large networks exploiting properties of wireless channels scales with the number of nodes n under various degrees of collaboration between terminals. When there is no cooperation and each terminal pair relies on long-range point-to-point communication, interference precludes simultaneous communication such that each pair only gets $1/n$ of the total bandwidth, and the total system capacity does not increase with n. Using packet multihop, communication can be confined to nearest neighbors, allowing more simultaneous transmission in the network, resulting in a total capacity scaling proportionally to $\sqrt{n}$ [16]. In contrast, by using more sophisticated cooperation techniques, it is possible to achieve a linear scaling of the network capacity [17] (and related [18][19][20][21]). The key theoretical idea is to have clusters of terminals form distributed antenna arrays to perform long range multiple-input-multiple-output (MIMO) transmission [22][23]. The cooperative MIMO transmission can mitigate interference, by joint processing of the received signals at each cluster, as well as perform beamforming for maximizing received power and range. These MIMO gains can be maximized by scaling the size of each cluster using hierarchical cooperation architecture to minimize overhead.

This theoretical result of linear system capacity increase with the number of users added presents a strong motivation for its implementation in a practical setting.

In order to operate in a cognitive mode, terminals have to be either allocated the open bands by the base station, or have to be able to sense the 'white space.' An individual terminal has limited ability of sensing a primary user operating in that area, as it always has a problem of being in a fade. Cooperation between a number of terminals can be used to greatly reduce the probabilities of misdetection and false alarms [24].

Wireless infrastructure. Technologies used for local communication, such as wireless LAN's are being deployed over wide, metropolitan-size areas. One interesting possibility is to shift from having the wired infrastructure to be necessary for any operation, to viewing it as a performance-boosting accelerator for a system that could also operate purely wirelessly. Such a perspective leads naturally to systems that can bridge gaps that occur because of disabled base stations.

A system that is based on solely a wireless connectivity, where both the terminals and the connectivity points can interoperate between different systems, besides its ability to provide efficient commercial communications, provides a network that can operate even in major disasters.

V. CONCLUSION

Today's wireless systems are limited by one hundred years of spectrum sharing, traditional narrowband radio architectures, limited cooperation within a system and between the system and the dependence on a wired infrastructure. Concept of cognitive and collaborative wireless communications, where vast amounts of spectrum are being used on a secondary basis through collaboration between the terminals, has a promise to dramatically expand the capacity of wireless systems.

ACKNOWLEDGMENT

Many of the ideas presented in this paper originated in discussions with author's colleagues, professors and students at the Berkeley Wireless Research Center and the Wireless Foundations Center at UC Berkeley. In particular, the author thanks Robert Brodersen, Jan Rabaey, Ali Niknejad, Paul Gray, Anant Sahai, Adam Wolisz, David Tse, Kannan Ramchandran, Danijela Cabric, Ahmad Bahai, and Howard Shelanski for contributions to these ideas. The sponsors of the Berkeley Wireless Research Center are also acknowledged.

REFERENCES

[1] J. Mitola III, "Cognitive Radio for Flexible Mobile Multimedia Communications," *Mobile Networks and Applications*, Vol. 6, No. 5, Sept. 2001.

[2] W.Krenik, A.Batra, "Cognitive radio techniques for wide area networks," *Proc. 42nd Design Automation Conference*, 13-17 June 2005 pp.409 – 412.

[3] D. Cabric, I. O'Donnell, M. S.-W. Chen, R.W. Brodersen, "Spectrum sharing radios," *IEEE Circuits and Systems Magazine*, June 2006.

[4] A. Akella, G. Judd, S. Seshan, and P. Steenkiste, "Self Management in Chaotic Wireless Deployments," MobiCom 05, August 28- September 2, 2005, Cologne, Germany

[5] N.Golmie, N. Chevrollier, O.Rebala "Bluetooth and WLAN coexistence: challenges and solutions" *IEEE Personal Communications*, Dec 2003

[6] X. Jing and D. Raychaudhuri, "Spectrum co-existence of IEEE 802.11b and 802.16a networks using CSC etiquette protocol," IEEE DySpan, 2005.

[7] S. Pollin, M. Ergen, M. Timmers, A. Dejonghe, L. Van der Perre, I.Moerman, F. Catthoor, and A. Bahai, \Distributed cognitive coexistence of 802.15.4 with 802.11," Crowncom, 2006.

[8] FCC, Revision of Part 15 of the Commission's Rules Regarding Ultra-Wideband Transmission Systems, First Report and Order, February 14, 2002.

[9] FCC, Unlicensed Operation in the TV Broadcast Bands, Notice of Proposed Rulemaking and Order, May 13, 2004.

[10] FCC, FCC Seeks Comment on Implementation of a Nationwide, Broadband, Interoperable Public Safety Network in the 700MHz Band, Notice of Proposed Rulemaking and Order, December 20, 2006.

[11] R. Gharpurey, S. Ayazian, "Feedforward Interference Cancellation in Narrow-Band Receivers," *2006 IEEE Dallas/CAS Workshop on*

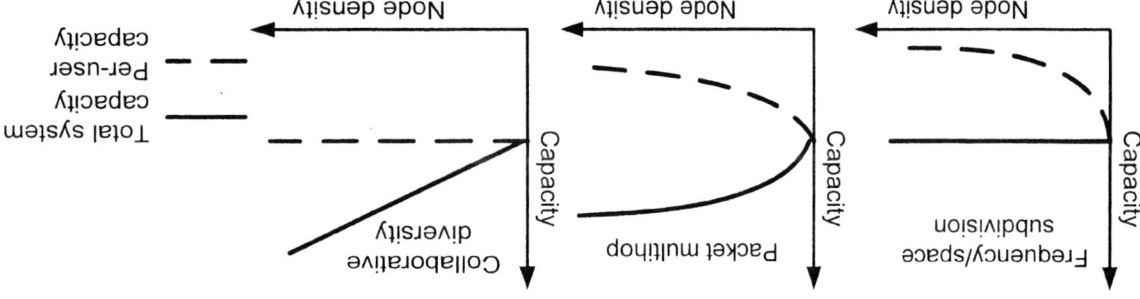

Figure 3: System capacity vs. node density as a function of degree of communication.

Design, Applications, Integration and Software, Oct. 2006 pp. 67 – 70.

[12] H. Darabi, "A Blocker Filtering Technique for Wireless Receivers," *International Solid-State Circuits Conference, ISSCC.2007, Digest of Technical Papers*, San Francisco, CA, Feb. 2007.

[13] F. Bruccoleri, E.A.M. Klumperink, B. Nauta, "Noise cancelling in wideband CMOS LNAs," *International Solid-State Circuits Conference, ISSCC.2002, Digest of Technical Papers*, pp. 406-407.

[14] W.-H. Chen, G. Liu, Z. Boos, A. M. Niknejad, "A Highly Linear Broadband CMOS LNA Employing Noise and Distortion Cancellation," to appear at IEEE RFIC 2007.

[15] J. R. Clark, W. Hsu, M. A. Abdelmoneum, C. Nguyen, "High-Q UHF micromechanical radial-contour mode disk resonators," *IEEE/ASME J. Microelectromechanical Systems*, vol. 14, no.6, 1298-1310, 2005.

[16] P. Gupta and P.R. Kumar, "The Capacity of Wireless Networks", *IEEE Trans. on Information Theory*, vol. 42, no.2, 2000, 388-404.

[17] A. Ozgur, O. Leveque and D. Tse, "Hierarchical Cooperation Achieves Optimal Capacity Scaling in Ad Hoc Networks", *IEEE Transactions on Information Theory*, vol 53, no. 10, pp. 3549 - 3572, October 2007.

[18] S.Aeron, V. Saligrama, "Wireless Ad hoc Networks: Strategies and Scaling Laws for the Fixed SNR Regime", Preprint, 2006.

[19] L.-L. Xie, P.R. Kumar, "A Network Information Theory for Wireless Communications: Scaling Laws and Optimal Operation", *IEEE Trans. on Information Theory*, vol. 50, no.5, pp. 748-767. 2004.

[20] L.-L. Xie, P.R.Kumar, "On the Path-Loss Attenuation Regime for Positive Cost and Linear Scaling of Transport Capacity in Wireless Networks", Preprint, 2006.

[21] A. Jovicic, S.R.Kulkarni and P.Viswanath, "Upper Bounds to Transport Capacity of Wireless Networks", *IEEE Trans. on Information Theory*, vol.50, no.11, pp.2555-2565. 2004.

[22] G.J. Foschini, "Layered Space-Time Architecture For Wireless Communication in a Fading Environment when Using Multi-Element Antennas", *AT&T Bell Labs. Tech. J.*, vol.1 no.2, pp.41-59, 1996.

[23] E. Telatar, "Capacity of Multi-Antenna Gaussian Channels" *European Trans. on Telecommunications*, ETT, vol.10, no.6, pp. 585-596, Nov. 1999.

[24] S.M. Mishra, A. Sahai, and R. Brodersen, "Cooperative sensing among cognitive radios," *Proceedings of IEEE ICC*, Istanbul, Turkey, Jun 2006.

Linearization of Highly-Efficient Monolithic Class E SiGe Power Amplifiers with Envelope-Tracking (ET) and Envelope-Elimination-and-Restoration (EER) at 900MHz

Donald Y.C. Lie, J.D. Popp*+, F. Wang*, D. Kimball*, and L.E. Larson*

Department of Electrical and Computer Engineering, Texas Tech University, Lubbock, TX 79409, USA
*Department of Electrical and Computer Engineering, University of California, San Diego (UCSD), La Jolla, CA
+ Orora Design Technologies, Inc., Redmond, WA 98052, USA
Email: Donald.Lie@ttu.edu

Abstract—**The linearization of highly efficient monolithic SiGe Class E power amplifiers (PAs) using both Envelope-Tracking (ET) and Envelope-Elimination-and-Restoration (EER) techniques has been studied at 900MHz. Without applying any linearization, the fully-integrated SiGe PAs achieve power-added efficiency (PAE) of 66% with no off-chip matching. The overall PAE of an ET-linearized PA system is 45% at an output power of 20dBm for an 881MHz EDGE (Enhanced Data Rate for GSM Evolution) modulated signal. The ET-linearized PAs pass the stringent EDGE transmit spectrum mask, but the EER-linearized PAs do not. The PAE of the ET system is expected to reach ~50% with further efficiency improvement on the envelope amplifier.**

I. INTRODUCTION

The power-added-efficiency (PAE) of a RF power amplifier (PA) is critical for the overall system performance, size, and reliability for portable transceiver products where power consumption is desired to be kept as low as possible. Switching-mode amplifiers (i.e., class D/E/S) can provide the highest possible PAE by operating the devices as switches to minimize the overlapping of current and voltage waveforms. A class E PA is easier for integration compared with a class F PA, and it is arguably the most efficient class of PAs because of the optimal switching conditions that minimize power dissipation [1-3]. In reality, non-idealities such as finite switching speed, switch resistive loss, passive component loss, device breakdown and voltage rail limitations, etc. have kept the PAE of the best Si-based RF class E PAs below 70% at 2 GHz and above [4-6]. The Class E PA operation requires both high peak voltage and current for the switching device, which seriously limits the safe operation region for deep-submicron CMOS Class E PAs [4].

With modern wireless communication systems utilizing more spectrally efficient and higher data rate modulation schemes, highly linear PAs are required to avoid the out-of-channel interference and distortion. For example, the EDGE (Enhanced Data Rate for GSM Evolution) waveform occupies 200KHz transmit channels within 880-910MHz with a moderate peak-to-average power ratio (PAR) of 3.3dB but it has stringent transmit (TX) spectral mask specs as GSM at -54dBc (400KHz), -60dBc (600KHz), and worst case rms EVM of 10%. Typically, the EDGE linearity requirement is achieved by using traditional current-mode PA (typologies (Class A/AB) and operating the amplifier several dB below its P_{1dB} compression point, which has

inevitably degraded its PAE significantly. Therefore, current commercial III-V EDGE PA modules can only achieve ~30% peak PAE at its maximum power because the PA is forced to operate in a back-off mode to keep it linear. Switched-mode PAs are intrinsically nonlinear in nature so they need to be linearized using techniques such as EER/ET for EDGE applications. The EER technique achieves this by providing constant amplitude phase-only signals to the Class E PA input, and separately drives the amplitude envelope modulation on the voltage supply of the PA. The ET technique also adds a percentage of the amplitude envelope modulation to the PA input. Classical ET-technique has been applied mainly to current-mode PAs. Our current research indicates that ET can be extended to switch-mode PAs and provide better linearization than the EER techniques. Fig. 1 shows a simplified block diagram of the ET PA linearization system used for this work. This paper details the implementation of highly-efficient monolithic SiGe class E RF PAs linearized using an ET technique to meet the linearity requirement for low-band EDGE modulated signals.

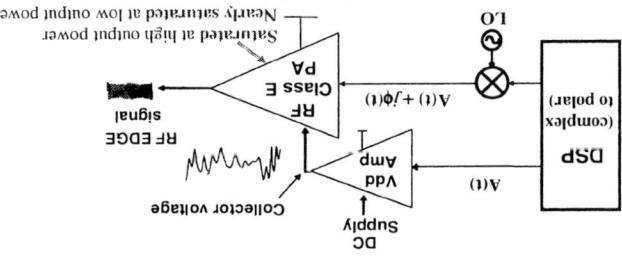

Fig. 1 A general ET PA linearization system diagram

II. RF CLASS E PA DESIGN AND RESULTS

A simplified schematic for our monolithic 1-stage Class-E PA for 900MHz operation is shown in Fig. 2.

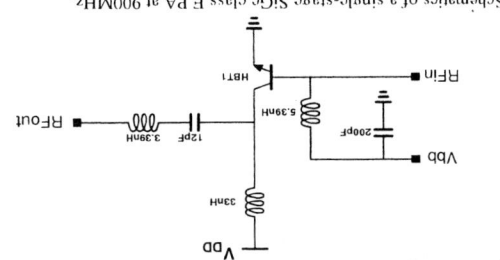

Fig. 2 Schematics of a single-stage SiGe class E PA at 900MHz.

III. ENVELOPE TRACKING AND EBR FOR EDGE

Both EBR and ET utilize dynamic power supply control schemes. Either the bias current and/or voltage is varied to control the instantaneous power. EBR uses a combination of a switching-mode PA and an envelope modulation circuit. Classical ET utilizes a switch-mode PA and a supply modulation circuit where the supply voltage V^{supply}_{dd} tracks the input AM envelope. In modern EBR/ET systems, the amplitude and phase signals are generated directly in the baseband domain and up-converted to RF. For the complex modulated signal, the complex baseband signal $s_{BB}(t)$ can be expressed with $I(t)$, $Q(t)$ or amplitude signal $A(t)$ and phase signal $\phi(t)$ as:

$$S_{BB}(t) = I(t) + j \cdot Q(t) = A(t) \cdot e^{j\phi(t)} \quad (1)$$

where $A(t) = \sqrt{I(t)^2 + Q(t)^2}$, $\phi(t) = e^{j\arctan(Q(t)/I(t))}$. Due to these nonlinear operations, bandwidth of $A(t)$ and $\phi(t)$ are much wider than that of baseband signal $s_{BB}(t)$ [6]. A block diagram of UCSD's open-loop ET system is shown in Fig. 6. Compared to EBR, ET can reduce the bandwidths of both envelope amplifier (i.e., V_{dd} amplifier) and the EBR-limiter for phase input [10].

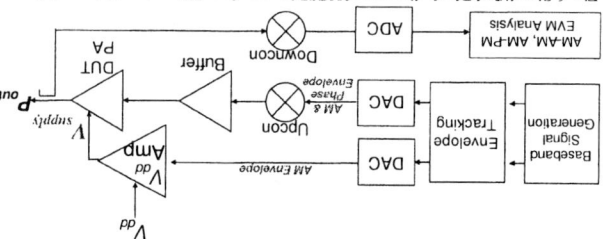

Fig. 6 Simplified Block diagram of UCSD's *open-loop* ET system (delay not shown; output feedback for analysis only)

The time delay mismatch between the RF input signal and voltage supply amplitude modulation in EBR/ET is a significant concern for EVM distortion. As concluded in [11], the ET technique applied to hard limiting PA's provides less sensitivity to delay mismatch than EBR. An ET system also relaxes the bandwidth requirement for the V_{dd} amplifier and for the RF path vs. EBR. ET system also provides higher gain at low output power than EBR since the device is nearly saturated at low output power (and saturated at high output power). An additional advantage of the PA operated with EDGE is that the moderate 3.3dB PAR and the linear P_{out} vs. V_{cc} and V_b responses (shown in Fig. 5) make it possible that digital pre-distortion is *not* needed to meet spectral mask and EVM requirements.

The simplified circuit schematics of the V_{dd} amplifier has been reported before [10,12,13], which uses a discrete linear op-amp and a switching power converter to provide amplitude

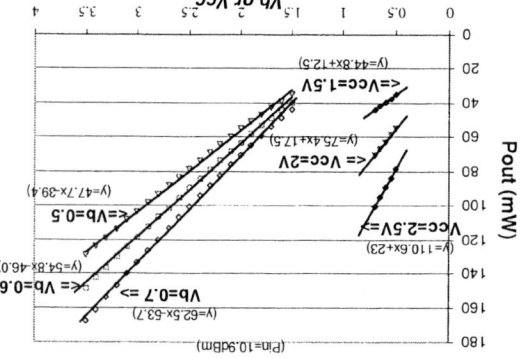

Fig. 5 Linear change in P_{out} (in mW) vs. supply V_{cc} and bias V_b (P_{in}=10.9dBm)

We purposely left the RF Choke (RFC) inductors off-chip because of the available low Q and large size for on-chip inductors at 900MHz. All the rest of the components are fully-integrated on-chip. The results in this paper are all obtained with fix-V_{bc} voltage biasing. The SiGe PA dies were bonded onto RF PC boards by bondwires for testing. A fabricated SiGe PA die is shown Fig. 3. The IBM 7HP technology high-f_T HBT device has a typical f_T/f_{max}~100/120GHz, BV_{CEO}=1.8V, and BV_{CBO}=6.4V while the high-breakdown HBT device has a typical f_T/f_{max}~27/57GHz, BV_{CEO}=4.2V, and BV_{CBO}=12.5V. Both HBT device options were investigated for in fabricated PAs. We designed these PAs with a voltage supply of 2.5-3.6V in mind, deliberately pushing above the BV_{CEO} limit of these HBTs. Both PAs designed using high-f_T and high-breakdown devices are tested. As expected, PAs designed with high-breakdown devices can withstand significant higher supply voltages and deliver higher output power P_{out}. No *off-chip* I/O matching for PAs is used here (measurement using off-chip tuning increased PAE further).

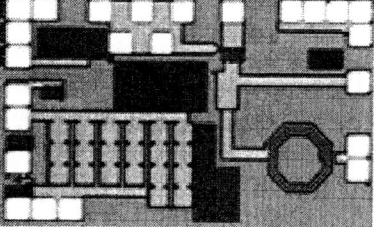

Fig. 3 A fabricated SiGe Class E PA micrograph (1.1mm x 1.7mm)

Single tone testing was first completed on the RF Class E PA. Fig. 4 shows the measured PA performance at 3.3V (72.5% CE, 65.6% PAE, and 22.5dBm P_{out}). Increases in the input power show the saturating nature of the Class E above P_{in}~5dBm.

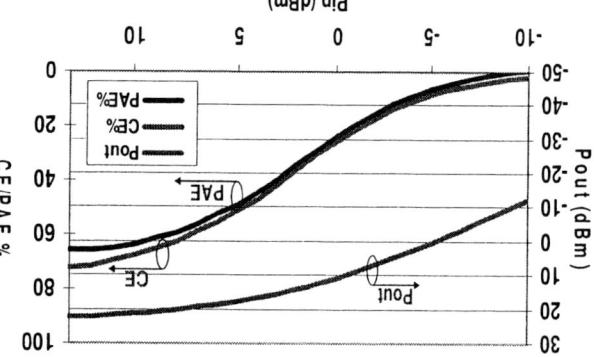

Fig. 4 Measured Class E Output Power/PAE vs. input power

Figure 5 shows a surprisingly *linear relationship between P_{out}* (in mW) vs. supply voltage V_{cc} and vs. base bias V_b. This is very different from the classical class E PA equations where a quadratic relationship between P_{out} vs. V_{cc} is expected. The reason is likely due to that the loaded Q of our PAs are only ~0.38 at 900MHz, significantly lower than the low-end validity limit of classic class E model of Q~1.78 [7]. The PA is more sensitive to changes in V_b than in V_{cc}, which is expected considering the exponential I-V behavior of the base-emitter diode. These linear relationships suggest that ET may be effective in linearizing this switch-mode PA, since to 1st-order the output power level can be *linearly* controlled by the base and collector bias voltages *independently* across wide ranges [8]. The device saturation voltage is ~0.8V as suggested from Fig. 5.

modulation envelope to the Class E PA voltage supply. Optimum PA system PAE is achieved by balancing the distribution of power supply current between the linear amplifier and the MOS switcher. It is critical to design a highly efficient V_{dd} amplifier, since the overall ET/EER system efficiency is a product of the PAE of the PA with that of the V_{dd} amplifier. In our current ET/EER system setup, the bandwidth of the envelope amplifier was purposely designed to be very wideband (~20MHz) to also accommodate the WLAN applications, and subsequently the amplifier only has an efficiency of 60-70% while generating the EDGE envelope. Work has been done to reduce the bandwidth of V_{dd} amplifier and therefore improves its PAE to ~80% range so that the overall PAE of the ET system is expected to improve to ~50% for the narrow-band EDGE signal.

IV. ET/EER RESULT OF THE LINEARIZED CLASS E PA IN EDGE

The measured RF Class E PA AM-AM performance for the RF output amplitude versus V_{supply} for EDGE amplitude modulated ET signal is shown in Fig. 7, without applying any ET/EER linearization. One can see the AM-AM relationship is very nonlinear and with considerable memory effects. Fig. 8 shows the AM-PM behavior for the class E PA, where the output phase envelope has large data variation, and it becomes very large at low input amplitude levels. It has been reported that for EDGE signal, the phase is not only dependent on the instantaneous amplitude but also a function of the history of the amplitude, causing some memory effects [14]. Figs. 9-10 show the measured AM-AM, AM-PM performances for the Class E PA linearized with the EER system, respectively (no pre-distortion). Great improvement on AM-AM linearity is accomplished; however, significant nonlinearity can still be observed at low input amplitude envelope level (<0.15 in Fig. 9), which is related to the low-end collector saturation voltage for the device and difficult to eliminate this nonlinearity in practice. The data variation for the EER-linearized PA in AM-PM curve is also very noticeably improved in Fig. 10. However, the AM-AM non-linearity apparently caused significant distortion as the best EER-linearized PA still failed the EDGE TX spectrum mask by several dB (-53dBc@400kHz and -62dBc@600kHz). Figs. 11-12 show the measured AM-AM, AM-PM behaviors for the ET-linearized class E PA system, respectively. Excellent AM-AM linearity is observed in Fig. 11, while the AM-PM behavior is reasonably good and similar to the case in Fig. 10 for EER-linearized PA. Note unlike CMOS PA, the linearity of our SiGe switch-mode PAs appears fairly insensitive to AM-PM variation for an EDGE modulated signal [15].

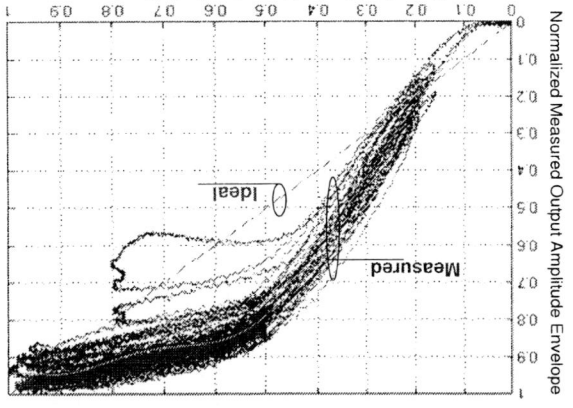

Fig. 7 Measured Class E SiGe PA E AM-AM characteristics (i.e. output amplitude vs. V_{supply}) without using any linearization.

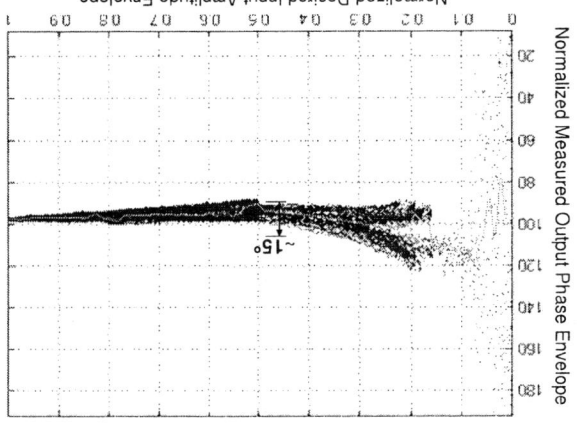

Fig. 8 Measured Class E SiGe PA AM-PM characteristics (i.e. output phase vs. V_{supply}) without using any linearization.

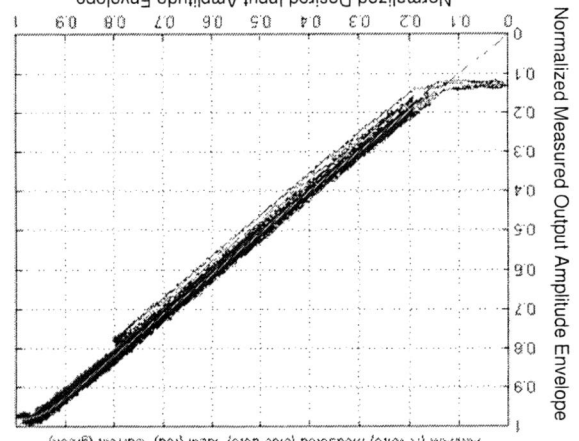

Fig. 9 Measured Class E SiGe PA AM-AM characteristics, PA linearized using the EER technique without pre-distortion.

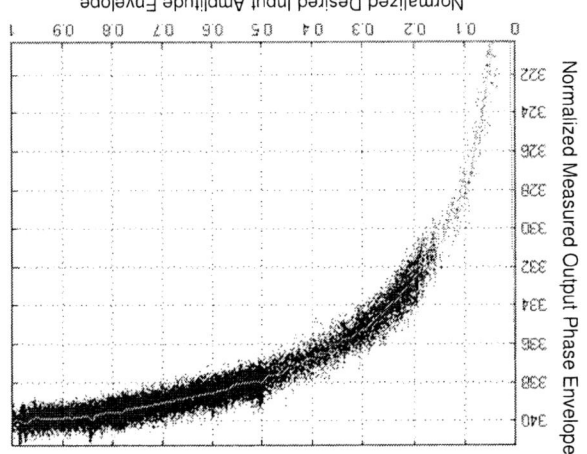

Fig. 10 Measured Class E SiGe PA AM-PM characteristics, PA linearized using the EER technique without pre-distortion.

Measurements of the Class E PA output spectrum and the ET Class E PA output spectrum are shown in Fig. 13 for EDGE signal (V_{cc}=3.0V, V_{vb}=0.65V, F_{in}=881MHz). It is clear that the nonlinearity of the Class E PA causes significant distortion when ET is not present. Measured results show a >14dB improvement

V. CONCLUSION

The linearization of highly efficient monolithic SiGe Class E PAs using both EER and ET techniques has been studied using an EDGE modulated signal. The ET-linearized PAS passed the stringent EDGE transmit spectrum mask without pre-distortion, while the EER-linearized PAs did not. Nonlinearity in the AM-AM characteristics present in the EER-linearized PAs (from saturation of the output power at low input power levels) suggests why ET is better than EER for this case. The overall PAE of the ET-linearized PA system is ~45% at 20dBm output for an 881MHz EDGE modulated signal. The PAE of the ET system is expected to reach ~50% with further efficiency improvement on the V_{dd} amplifier and with off-chip matching.

ACKNOWLEDGMENT

We thank Prof. P. Asbeck and Mr. P. Draxler at UCSD for valuable discussions. We thank DoD for support and IBM for IC fabrication.

REFERENCES

[1] G. D. Ewing, "High-Efficiency Radio-frequency Power Amplifier", PhD thesis, Oregon State University, Corvallis, Oregon, June (1964)

[2] N.O. Sokal, A.D. Sokal, "Class E-a new Class of high efficiency tuned single-ended switching power amplifiers", IEEE J Solid-State Circuits, vol. 10, pp. 168-176, (1975)

[3] F.H. Raab, "Idealized operation of the Class E tuned power amplifier," IEEE Trans. Circuits Syst, vol. 24, pp. 725-35, (1977)

[4] K. Tsai and P. Gray, "A 1.9-GHz, 1-W CMOS Class-E Power Amplifier for Wireless Communications," IEEE Journal of Solid-State Circuits, vol. 34, no. 7, pp. 962-970, (1999)

[5] C. Yoo and Q. Huang, "A common-gate switched 0.9-W Class-E power amplifier with 41% PAE in 0.25μm CMOS," IEEE Journal of Solid-State Circuits, vol. 36, pp. 823-30, (2001)

[6] "The Limitations in Applying Analytic Design Equations for Optimal Class E RF Power Amplifiers Design", D.Y.C. Lie, J. Lopp, J. Rowland, H. H. Ng, and A. Yang, Proc. Tech. Dig., IEEE Int'l Symp. on VLSI Design, Automation and Test (VLSI-TSA-DAT), pp. 161-164, Hsin-Chu, Taiwan, April 27-29 (2005)

[7] N.O. Sokal, "Class-E Switching-Mode High-Efficiency Tuned RF/Microwave Power Amplifier: Improved design equations", IEEE MTT-S, pp. 779-782, (2000)

[8] D.Y.C. Lie and J.D. Popp, "A novel way of maximizing the output power efficiency for Switch-mode RF Power Amplifiers", US Patent claims approved, Nov. 2006, Waiting for U.S. patent No.

[9] B.P.Lathi, Modern Digital and Analog Communication Systems, Oxford University Press, (1998)

[10] F. Wang, D. Kimball, J. Popp, A. Yang, D.Y.C. Lie, P. Asbeck and L.E. Larson, "Design of Wide Bandwidth Hybrid Envelope Elimination and Restoration Power Amplifiers for Wideband OFDM Applications", IEEE Trans. Microwave Theory Tech, 54, 12, pp. 4086-4099 (2006)

[11] "Design of Wide-Bandwidth Envelope-Tracking Power Amplifiers for OFDM Applications, ", F. Wang, A. Yang, D. Kimball, L.E. Larson and P. M. Asbeck, IEEE Trans. Microwave Theory and Techniques, pp. 1244-1255, 53, 4, (2005)

[12] J. D. Popp, D.Y.C. Lie, F. Wang, D. Kimball, P. Asbeck and L.E. Larson, "A Fully-Integrated Highly-Efficient RF Class E SiGe Power Amplifier with an Envelope-Tracking Technique for EDGE Applications", Dig. IEEE Radio and Wireless Symposium (RWS 2006), p.p. 231-234, San Diego, Jan. (2006)

[13] "A Monolithic High-Efficiency 2.4 GHz 20 dBm SiGe BiCMOS Envelope Tracking OFDM Power Amplifier", F. Wang, D. Kimball, D.Y.C. Lie, P. Asbeck and L.E. Larson, IEEE J. Solid-State Circuits, 42, 6, pp. 1271-1281 (2007)

[14] "EDGE transmitter with commercial GSM power amplifier using polar modulation with memory predistortion", G. Seegerer and G. Ulbricht, IEEE MTT-S International Microwave Symposium Digest, June 12-17, pp. 1553-1556 (2005)

[15] "A 5-GHz 20-dBm Power Amplifier With Digitally Assisted AM-PM Correction in a 90-nm CMOS Process", Y. Palaskas, S.S. Taylor, S. Pellerano, I. Rippke; R. Bishop; A. Ravi; H. Lakdawala; K. Soumyanath, IEEE J. Solid-State Circuits, 41, 8, pp. 1757-1763 (2006)

at the 400KHz and 600KHz offsets using ET on the Class E PA with EDGE modulation signal.

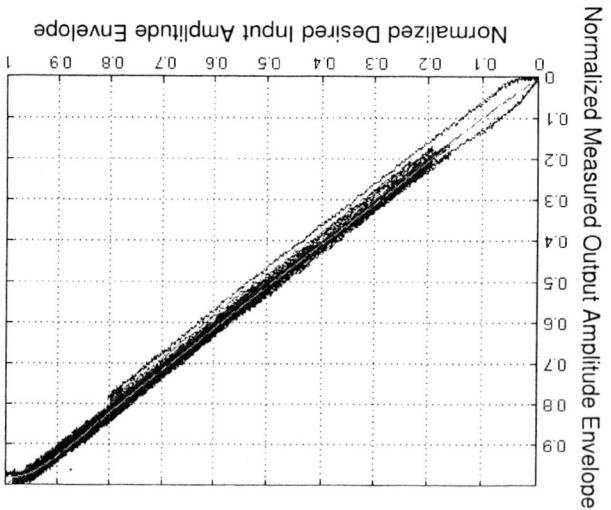

Fig. 11 Measured Class E SiGe PA AM-AM characteristics. PA linearized using the ET technique without pre-distortion.

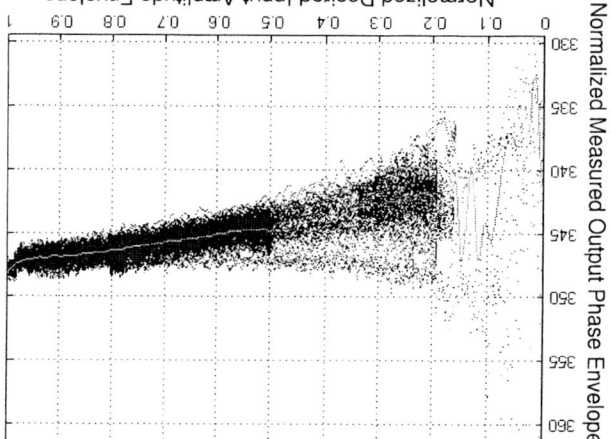

Fig. 12 Measured Class E SiGe PA AM-PM characteristics. PA linearized using the ET technique without pre-distortion.

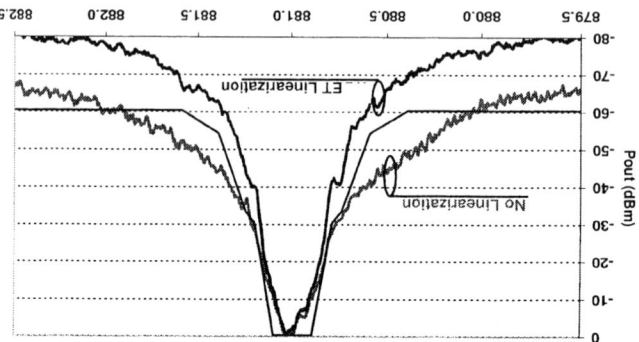

Fig. 13 Class E SiGe PA EDGE TX output spectrum before and after ET linearization.

DIGITAL FIR FILTER OPTIMIZATION USING TOGGLE-BASED POWER ESTIMATION TOOLS

Cristian M. Albina, Günther Hackl

GME mbH, 82008 Unterhaching, Germany

ABSTRACT

In this paper one method of optimizing a digital finite impulse response (FIR) filter has been illustrated. The advantages and disadvantages of several architectures and of the circuit modeling were presented using a standard toggle-based method for the circuit power estimation, gate-level simulations and synthesis. We showed that we can achieve a significant power reduction from the beginning by carefully selecting the right architecture and optimizing the VHDL code description of the module. The analysis was made based on the unity delay model and not on the physical extracted layout for a 150nm technology but the method can be used for other technologies as well.

Index Terms—FIR, VHDL, RTL, Power, Synthesis

1. INTRODUCTION

Over the years, the state-of-the-art technologies pushed the VLSI chips to higher clock speed and packing density. The trends for the coming years are defined already by the International Technology Roadmap for Semiconductors (DSP) continue to expand, driven by trends such as the increased use of video, audio and still images. Many of these applications combine the need for significant DSP processing with cost sensitivity, creating demand for high-performance, low-cost DSP solutions. Because the power consumption has become an important factor in the design process of a chip due to the limited lifetime of the battery fast and accurate power estimation tools are needed for each level of the circuit in order to ensure that the energy and the design constraints are met. There are already several power estimation tools [2], [3], [4] which can help the designer to this task but the more detailed the simulation is the longer the time and the bigger the amount of data to be simulated. In our paper we'll use the standard power estimation method for the CMOS circuits using gate-level and RTL (Register Transfer Level) based on unit delay model simulations combined with Synopsys [5] synthesis tool to show the effect of the RTL architecture and of the VHDL description [6] on the final area of the module as well as on the total power, solution which offers the advantage of small designs

with lowest power consumption and which are highly correlated to the final layout.

2. DIGITAL FIR FILTERS

The basic structure of a Finite Impulse Response (FIR) filter is shown in Figure 1. The multipliers and the adders form the heart of a FIR filter. The input data passes to the multiplier and then to the adder with interleaving delay elements.

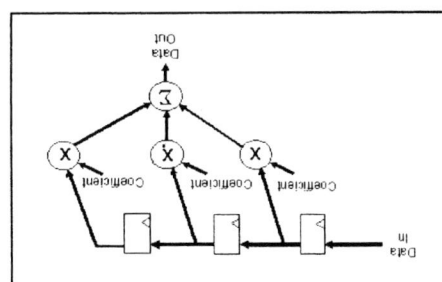

Figure 1. Finite Impulse Filter Functional Diagram

The direct-form and transpose-form structures are most commonly used to implement FIR filters. For certain special filters, recursive implementations require less computation. Lattice and cascade structures are occasionally also used. It is finite because its response to an impulse ultimately settles to zero. This is in contrast to infinite impulse response filters which have internal feedback and may continue to respond indefinitely. The difference equation which defines how the input signal is related to the output signal is:

$$y(n) = \sum_{k=0}^{M-1} h_k \times x(n-k) \qquad (1)$$

where M is the filter order, x(n) is the input signal, y(n) is the output signal and h_k are the filter coefficients. The FIR transfer function obtained after applying a Kronecker delta impulse as an input signal contains M poles for z = 0. Since all poles are at the origin, all poles are located within the unit circle of the z-plane; therefore all FIR filters are stable.

3. IMPLEMENTED DIGITAL FIR STRUCTURES

We'll start to design the filter by selecting the coefficients based on the circuit specifications. Most of the time filter specifications refer to the frequency response of the filter. We used in our paper a 29th order single channel FIR (30 coefficients) with a system clock frequency of 100 MHz and for the evaluation we used the input and output sample rate equal to 100MHz/32 instead of 30 due to the better implementation ratio between the system frequency and the sampling rate. The filter pass-band is 1 MHz and the stop-band is 1.2 MHz at which the filter attenuation is - 60 dB. The whole module was implemented using a 150 nm CMOS technology models and V_{DD} 1.5 V. The synthesis parameters are: clock period 10 ns, critical path and area analysis, use enclosed wire load model, and use clock gating option.

3.1. Direct implementation based on the C++ code

In most cases the digital designer starts his implementation based on the direct behavioral description delivered by the system concept engineer. This description is usually written in C++ language and has no hardware related physical parameters included. The filter architecture is illustrated in Figure 2 and contains a shift register (SR), a coefficient ROM memory and a multiply and accumulate unit (MAU).

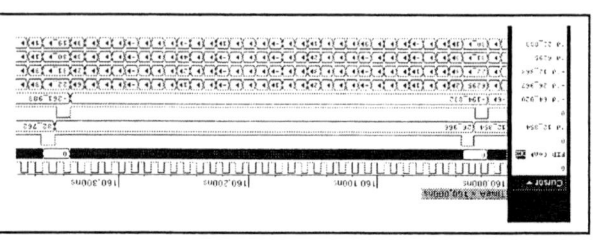

Figure 2. FIR architecture based on direct C++ code description

First we'll run the gate-level and the RTL simulations like in Figure 3.

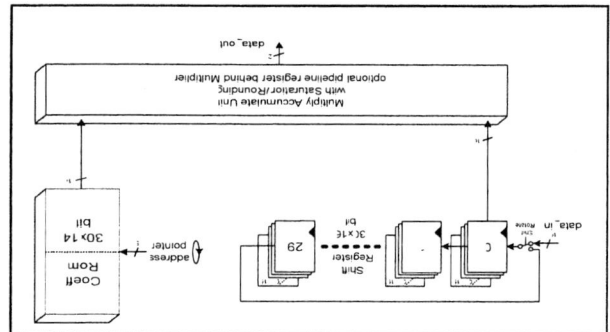

Figure 3. FIR simulation results

Analyzing the toggle activity report we'll get the following power consumption summary for this implementation.

Table 1. Direct FIR Power Consumption

Direct FIR Power Consumption					
Hierarchy Level	Switch Power (mW)	Int. Power (mW)	Leak Power (µW)	Total Power (mW)	Power %
FIR	6.649	10.61	15.34	17.279	100
Top level FIR logic and ROM	2.810	4.273	2.869	7.594	42.4
Adder inside MAU	0.641	0.309	0.147	1.239	12.5
Multiplier inside MAU	3.186	4.596	3.452	7.785	45.1

3.2. Improved implementation using a Read Decoder

In order to reduce the module power consumption we'll have to try to minimize the number of cycles necessary to read the filter coefficients out of the ROM memory.

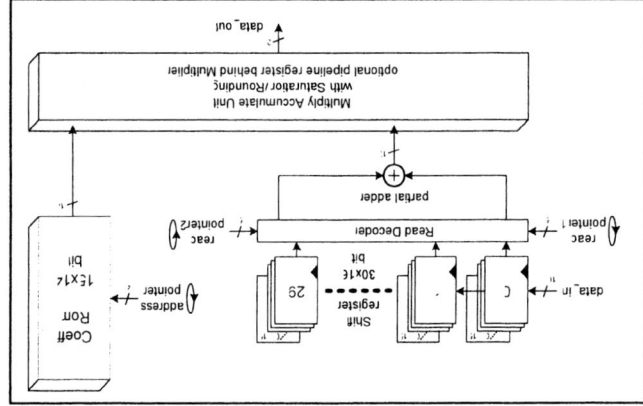

Figure 4. FIR architecture using a Read Decoder

One way to achieve this is by carefully choosing the coefficients and their position inside the memory in such a way that the whole process takes only half of the time like in the architecture illustrated in Figure 4 and by using a read decoder register. Redoing the RTL simulations, synthesis and gate-level simulations like in Figure 5 we obtain the following results for the module power consumption:

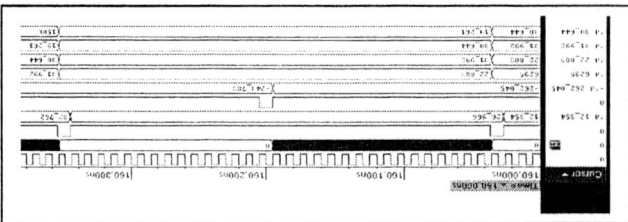

Figure 5. Improved FIR simulation results

We can notice the reduction of the filter convolution activity on the data bus, second signal in the waveform diagram. Furthermore there is an additional switching and internal power reduction inside the MAU (multiplier and adder) and on the top level due to the fact that the implementation needs less digital logic on the top level, most of the operations being already done by the read-decoder.

Table 2. Improved FIR Power Consumption

Hierarchy Level	Switch Power (mW)	Int Power (mW)	Leak Power (µW)	Total Power (mW)	Power %
FIR	3.495	4.543	15.72	8.054	100
Top level FIR logic and ROM	0.630	0.753	0.695	1.368	17.4
Adder inside MAU	0.065	0.242	0.338	0.307	3.8
Multiplier inside MAU	2.683	3.365	3.277	6.051	75.1
Partial adder	0.060	0.176	0.414	0.264	2.9
Read decoder	0.057	0.006	0.018	0.063	0.8

3.3. Implementation with a Read/Write Decoder

Several architectures and algorithms were discussed in the past years in the [7]-[9]. For the further power reduction we can use an additional write decoder combined with a special read/write arithmetic pointer which reduces furthermore the toggle activity. Such architecture is proposed in the Figure 6. In contrast to the previous implementations, the additional Write Decoder selects only one address location of the Data Buffer and writes the new sample into it. This mechanism avoids the complete rotation of the registers like it happens in the previous shift register implementations and therefore the amount of toggles get reduced. Once the new incoming data is stored, the write pointer is incremented and points now to the next address location of the Data Ring Buffer.

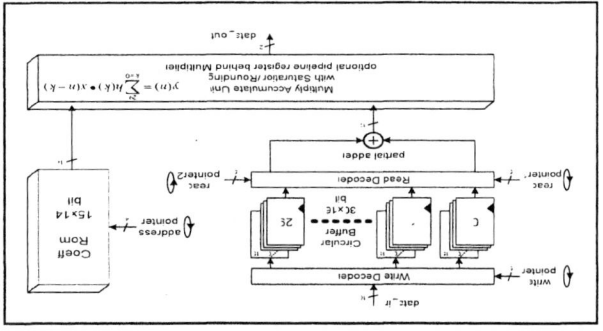

Figure 6. FIR architecture using a Read/Write Decoder.

A more detailed architecture of the Read/Write Arithmetic Pointer concept is illustrated in Figure 7.

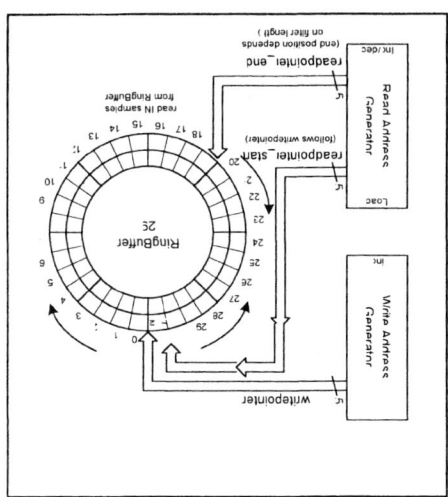

Figure 7. Read/Write arithmetic pointer architecture

The multiply and accumulate unit (MAU) was implemented based on the following architecture:

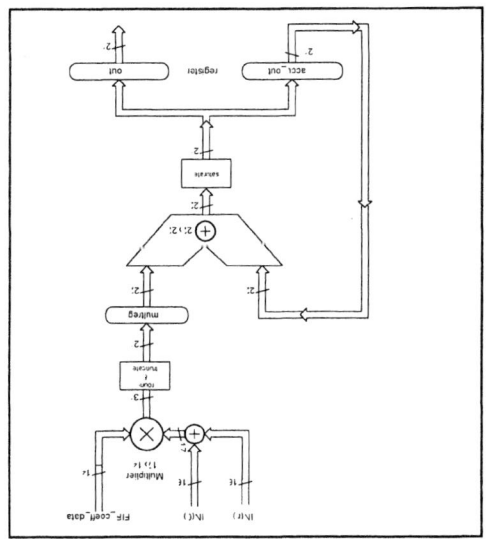

Figure 8. Multiply accumulate unit architecture

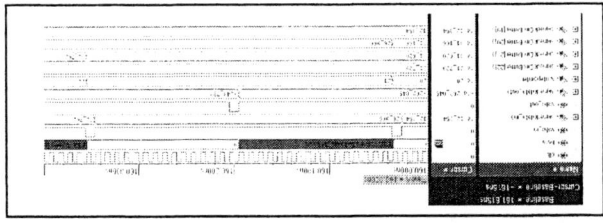

Figure 9. Read/Write FIR simulation results

Based on the simulation results from the Figure 8 we've got the following results for the module power consumption:

Table 3. Read/Write Pointer FIR Power Consumption

Hierarchy Level	Switch Power (mW)	Int Power (mW)	Leak Power (µW)	Total Power (mW)	Power %
FIR	3.007	4.189	17.69	7.214	100
Top level FIR logic and ROM	0.583	0.673	1.358	1.225	17.9
Adder inside MAU	0.041	0.153	0.253	0.195	2.7
Multiplier inside MAU	2.293	3.143	3.277	5.44	75.4
Circular Buffer read decoder	0.020	0.004	0.009	0.024	0.3
Circular Buffer write decoder	0.001	0.002	0.009	0.003	0.1
Circular Buffer 5 bit inc/dec	0.006	0.025	0.052	0.032	0.5
Read decoder	0.046	0.054	0.014	0.051	0.7
Partial adder	0.048	0.126	0.194	0.175	2.4

4. CONCLUSIONS

In the presented architectures we used SR (shift register) fully (32bit) or half (16 bit) addressed, CG (clock gating), CB (circular buffer) and parallel processing of the coefficients using pipe-line structures. By carefully optimizing the architecture we were able to reduce considerably the power consumption up to 60%.

Table 4. Comparison of different FIR architectures

Architecture	Module Area (mm²)	Toggle Count (Mills Tc)	Power (mW)	Details
FIR A	0.050	64.2	17.279	SR, full
FIR B	0.054	33.7	8.054	SR, half, CG pipe
FIR C	0.057	30.2	7.214	CB, half, CG pipe

Compared to other methods [10], we achieved the power reduction only through optimal addressing of coefficients and through the usage of the ring buffer instead of improving the samples computation or by adding pipeline registers. This method is suitable for programmable and symmetric adaptive filters. There are already several methods for fixed filter design which can lead to lower power consumption by making the design multiplierless and apply the common subexpression elimination to share the operators. This filter was already implemented using a 150 nm CMOS technology and the measurements done on the final silicon confirmed the simulation results for the main filter parameters.

There is a lot of work going around trying to find even better architectures and better coefficient quantization methods to improve furthermore the overall power consumption combined in parallel with the usage of modern submicron technologies. The general tendency is towards low power modules combined with intelligent algorithms for easy VLSI implementation.

5. REFERENCES

[1] International Technology Roadmap for Semiconductors 2004 Edition, http://public.itrs.net.

[2] Design Power Tools, http://www.synopsys.com/products/solutions/galaxy/power/power.html.

[3] M. Karunaratne, C. Ranasinghe, A. Sagahyroon, "A dynamic switching activity generation technique for power analysis of electronic circuits," Circuit and Systems, 2005, 48th Midwest Symposium on, Vol.2, pp. 1884-1887, 2005.

[4] Raghunathan, A.; Dey, S.; Jha, N.K. "High-level macro-modeling and estimation techniques for switching activity and power consumption", Very Large Scale Integration (VLSI) Systems, IEEE Transactions on, Volume 11, Issue 4, pp. 538 – 557, 2003.

[5] Synopsys, Power Compiler Reference Manual, http://www.synopsys.com/products/power/power-ds.pdf.

[6] Cadence NC-VHDL Simulator, http://www.cadence.com/products/functional_ver/nc-vhdl/index.aspx.

[7] P. K. Merakos, K. Masselos, K. Stouraitis, "Optimisation techniques for reducing global bus switching activity in realisations of sum-of-products computations in DSP systems," Circuits, Devices and Systems, IEEE Proceedings-Volume 150, Issue 1, pp. 16-25, 2003.

[8] Jongsun Park et al., "High performance and low power FIR filter design based on sharing multiplication", Low Power Electronics and Design, Proceedings of the 2002 International Symposium on, ISLPED '02, pp. 295 – 300, 2002.

[9] S.G. Chandar, M. Mehendale, R. Govindarajan, " Area and power reduction of embedded DSP systems using instruction compression and reconfigurable encoding", Computer Aided Design, 2001, ICCAD 2001, IEEE/ACM International Conference on, pp. 631-634, 2001.

[10] N. Nedovic, V.G. Oklobdzija, M. Leung, "FIR Filter for Adaptive Equalization in PRML Read Channels", The 5th World Multi-Conference On Systemics, Cybernetics and Informatics SCI, 2001.

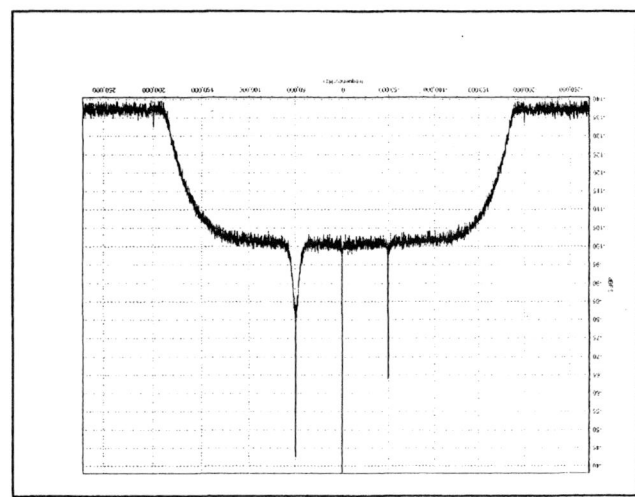

Figure 10. Read/Write FIR measured results

Analysis of Third-order Intermodulation in Receiver Down-Converter Employing Multiband Feedback

Jungbwan Han and Ranjit Gharpurey

Abstract—This paper presents an analysis of the third-order intermodulation characteristics of a recently reported receiver down-converter with multiband feedback and current reuse. This design not only achieves very a high gain with a low-current requirement but also provides an inherent cancellation mechanism for enhancing the linearity of the topology. By employing the signal feedback scheme, the inherent second-order nonlinearity of the tail current sources of differential pairs that are used to provide gain at baseband can be used to cancel the third-order intermodulation distortion resulting from the same devices. Circuit simulation using 0.13-μm RF CMOS process shows an IIP3 improvement of 4 dB in comparison to its normal cascaded counterpart.

Index Terms—CMOS, dual-band utilization, feedback, linearity improvement, receiver down-converter, third-order nonlinearity, third-order intermodulation (IM3) cancellation

I. INTRODUCTION

IN recent decades, low-cost and low-power wireless transceivers utilizing standard CMOS technology have attracted widespread attention for wireless personal area network (WPAN) applications in the 900 MHz and 2.4 GHz ISM bands, such as Zigbee and Bluetooth. The dynamic range requirements of these standards are modest in comparison to those of transceivers for other cellular applications; however, the requirement for low power is particularly critical. The ISM band is shared by several unlicensed wireless systems. Thus the linearity of transceivers must be considered to ensure that the performance requirements are satisfied.

We recently reported low-power high-gain receiver down-converters that utilize multiband feedback, that are well-suited for the above mentioned standards [1]. As shown in Fig. 1, by using a signal feedback path from the mixer output to the input transconductance stage, current source devices (M_{3a-b}) used for biasing input devices M_{1a-b} and M_{2a-b} are reused for baseband amplification. This approach demonstrates that by sharing the bias current for mixing and baseband amplification, the achievable gain is greatly enhanced without an increase in

This work was partially supported by the National Science Foundation (NSF) under grant number ECS-0602621

Jungbwan Han and Ranjit Gharpurey are with the Department of Electrical & Computer Engineering, University of Texas at Austin, TX 78712 USA (e-mail: jbhan@mail.utexas.edu and ranjit@mail.utexas.edu)

the power consumption and an unacceptable degradation of the dynamic range. In addition, input transconductance stages simultaneously operate in a single-ended mode at RF and in a differential mode at baseband to ensure stability without the requirement for sharp filters and to improve second-order linearity [1]. In this paper we analyze the third-order intermodulation (IM3) behavior of this topology.

In the case of a typical cascaded system, the overall linearity can usually be optimized by a trade-off between the design parameters of each block. In our feedback scheme, the down-converted baseband signals are reapplied to the input, which introduces the possibility that the nonlinearity in the input stage will generate additional beat products of the IF and RF signals. As shown below, these additional terms can be used to our advantage, such that the reported design (Fig. 1) can be designed for even better linearity than a conventional cascaded down-converter. Specifically, the nonlinearity of the differential pairs used as the RF input transconductance stage interacts with the inherent second-order nonlinearity of current source devices which can be used for canceling the third-order intermodulation of the tail current devices.

Nonlinearity cancellation has been studied in other contexts for example [2] proposed a linearization technique for common-source RF input transconductors by using the second-order nonlinearity of current mirror circuitry. In this

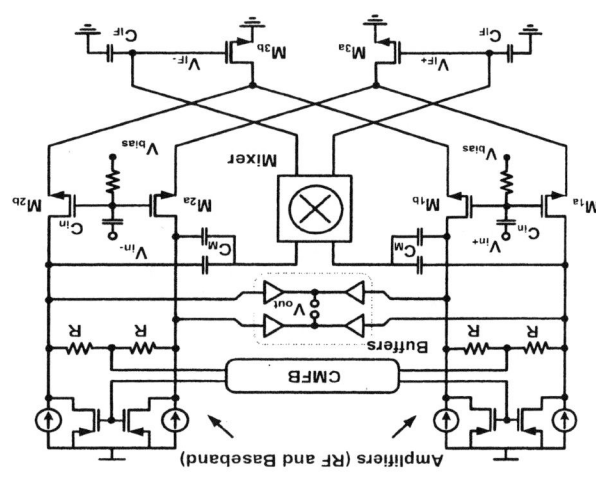

Fig. 1. Schematic representation of receiver down-converter [1]

paper, we examine the inherent properties of the topology of [1], instead of introducing circuits around a given topology for the purpose of cancellation.

II. THIRD-ORDER INTERMODULATION OF CASCADED RECEIVER DOWN-CONVERTER

Let us first consider a cascade of an RF differential amplifier, a mixer, and a baseband amplifier, as shown in Fig. 2. This can be recognized as the "opened" version of the design of Fig. 1. We assume that the mixer switches, resistors, and capacitors are perfectly linear and the input stage is the only source of nonlinearity. The output current of differential pair (Fig. 3) for square-law devices is given by [3]

$$i_{d1} - i_{d2} = \alpha \cdot v_{id} \cdot \sqrt{I_B - \gamma \cdot \frac{v_{id}^2}{2}} \qquad (1)$$

Following [3], the factors α and γ denote $\sqrt{k_n'(W/L)}$ and $k_n'(W/L)/4$, respectively. I_B is the tail current of the differential pair. In the design of Fig. 2, the total RF output current $i_{od} = 2 \cdot \alpha \cdot v_{id} \cdot \sqrt{I_B - \gamma \cdot \frac{v_{id}^2}{2}}$ flows into the mixer and is down-converted to IF. The differential voltage at the output of mixer is given by

$$2v_{IF} = v_{IF+} - v_{IF-} = G \cdot i_{od} \qquad (2)$$

where G is the product of the current conversion gain and load resistor R_L. Assuming ideal square-wave switching at the mixer, G is given by $(2/\pi) \cdot R_L$. If we express the nonlinearity of the following the baseband amplifier as the power series $i_{OIF} = \beta_1 v_{IF} + \beta_2 v_{IF}^2 + \beta_3 v_{IF}^3 + \cdots$, the overall nonlinearity at the output of baseband amplifier is given by

$$i_{OIF} = 2\beta_1 \left(G\alpha \cdot v_{id} \cdot \sqrt{I_B - \gamma \cdot v_{id}^2}\right)$$
$$+ 2\beta_3 \left(G\alpha \cdot v_{id} \cdot \sqrt{I_B - \gamma \cdot v_{id}^2}\right)^3 \qquad (3)$$

By employing a Taylor-series expansion, Eq. (3) can be re-written as

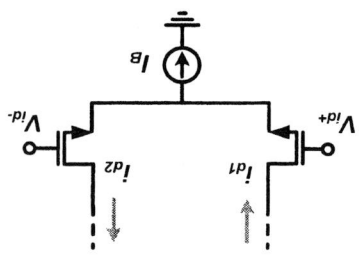

Fig. 3. Transconductance of differential-pair

$$i_{OIF} = \left(2\beta_1 G\alpha\sqrt{I_B}\right) \cdot v_{id}$$
$$+ \left(2\beta_3 G^3\alpha^3 I_B^{3/2} - \frac{\beta_1 G\alpha\gamma}{\sqrt{I_B}}\right) \cdot v_{id}^3 + \cdots \qquad (4)$$

$$\cong \left(2\beta_1 G\alpha\sqrt{I_B}\right) \cdot v_{id} + \left(2\beta_3 G^3\alpha^3 I_B^{3/2}\right) \cdot v_{id}^3 \qquad (5)$$

From Eq. (5), the dominant third-order nonlinearity is thus seen to arise from the third-order term β_3 of the baseband amplifier since the signal at its input is amplified by the conversion gain of the mixer. In the case of a cascaded down-converter, the IM3 product can be reduced by lowering the gain of the first stage, which leads to an increase in noise floor.

III. THIRD-ORDER INTERMODULATION DISTORTION OF FEEDBACK-BASED RECEIVER DOWN-CONVERTER

A similar derivation for the IM3 products is presented below for the simplified feedback-based down-converter shown in Fig. 4, which demonstrates the above mentioned cancellation mechanism. A detailed analysis of the reuse stage of the current source is presented below by assuming square-law devices in the input differential pairs. The analysis is not exact for short-channel devices. However it illustrates the cancellation principle.

As seen in Fig. 4, the down-converted signals are applied to the gates of tail current-source devices of the input differential

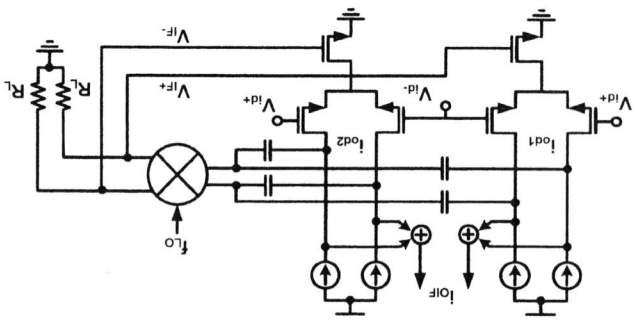

Fig. 2. Conventional cascaded down-converter consisting of different-pair, mixer and baseband amplifier stages

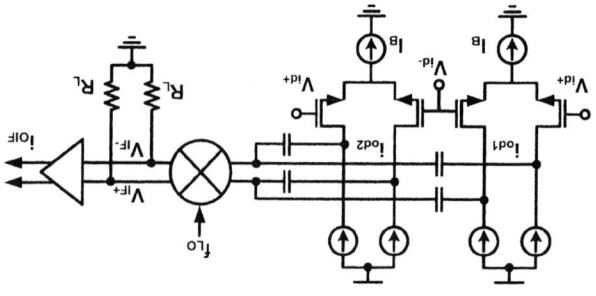

Fig. 4. Simplified feedback-based down-converter with multiband feedback

amplifiers for IF amplification. Due to this feedback connection, Eq. (1) needs to be modified to include the current variation caused by v_{IF}. Hence, for this case,

$$i_{od1} = \alpha \cdot v_{id} \cdot \sqrt{I_B + f\left(-v_{IF}\right) - \gamma \cdot v_{id}^2} \qquad (6)$$

$$i_{od2} = \alpha \cdot v_{id} \cdot \sqrt{I_B + f\left(v_{IF}\right) - \gamma \cdot v_{id}^2} \qquad (7)$$

where $f\left(v_{IF}\right) = \beta_1 v_{IF} + \beta_2 v_{IF}^2 + \beta_3 v_{IF}^3 + \cdots$ models the nonlinear small-signal dependence of the tail current on the IF voltage at the output of the mixer. For proper comparison, the polynomial coefficients equivalent to those of the baseband current source devices are assumed to have nonlinear voltage v_{IF} in the baseband amplifier in Fig. 2, namely, β_1, β_2, and β_3. The voltage v_{IF} in Eqs. (6) and (7) can be observed to satisfy the equation

$2v_{IF} = G \cdot (i_{od1} + i_{od2})$. We thus have

where $f(v_{IF})$ is the current variation in the tail current devices, including their nonlinearity. Eq. 7 yields

$$2v_{IF} = G\alpha v_{id}\left[\sqrt{I_B}\sqrt{1+\frac{f\left(v_{IF}\right)-\gamma v_{id}^2}{I_B}} + \sqrt{I_B}\sqrt{1+\frac{f\left(-v_{IF}\right)-\gamma v_{id}^2}{I_B}}\right] \qquad (8)$$

If $f(v_{IF})$ is assumed to be given by $\beta_1 v_{IF}+\beta_2 v_{IF}^2+\beta_3 v_{IF}^3$, then using this approximation in (8) and employing a Taylor-series expansion, we get

$$2v_{IF} = G\alpha v_{id}\sqrt{I_B}\left[2 + \left(\frac{\beta_2}{I_B} - \frac{\beta_1^2}{4I_B}\right)v_{id}^2 + \frac{\gamma\beta_2 v_{IF}^2 - 2\gamma}{2I_B}v_{id}^2 + \cdots\right] \qquad (9)$$

Further using the approximation $v_{IF} \cong c_1 v_{id} + c_3 v_{id}^3$, and solving for c_1 and c_3, we obtain

$$c_1 = G\alpha\sqrt{I_B} \qquad (10)$$

$$c_3 = \frac{G^3\alpha^3\gamma}{2}\sqrt{I_B}\left(\beta_2 - \frac{\beta_1^2}{4I_B}\right) - \frac{G\alpha\gamma}{2\sqrt{I_B}} \qquad (11)$$

Since the differential IF output current is given by

$i_{OIF} = 2\beta_1 v_{IF} + 2\beta_3 v_{IF}^3$, we have

$$i_{OIF} \cong \left(2G\alpha\beta_1\sqrt{I_B}\right)V_{id} + \left[G^3\alpha^3\beta_1\sqrt{I_B}\left(\beta_2 - \frac{\beta_1^2}{4I_B}\right) + 2G^3\alpha^3\beta_3 I_B^{3/2}\right]V_{id}^3 \qquad (12)$$

By comparing Eq. (5) with Eq. (12), it can be observed that an additional term $G^3\alpha^3\beta_1\sqrt{I_B}\left(\beta_2 - \beta_1^2/4I_B\right)$ contributes to the third-order nonlinear distortion. This term arises due to nonlinearity of the input differential pair that appears in response to the modulation of the tail current of the differential pair.

Typically in MOSFETs biased in saturation, β_2 is a positive quantity and β_3 is negative [4]. Thus, the overall third-order nonlinear characteristics can be adjusted by controlling the second-order nonlinear coefficient β_2. In the case that $\beta_2 = \beta_1^2/4I_B$, the IM3 product of the feedback-based down-converter becomes equal to that of its cascaded counterpart. By increasing β_2, even further, the overall IM3 performance is further improved. In this case the second-order nonlinearity of the tail devices is seen to cancel the IM3 product generated by these devices.

In addition to the inherent second-order nonlinearity of the current source devices, if required, an additional second-order term can also be introduced into the tail current source by means of a dedicated circuit with a variable output. As depicted in Fig. 5, a simple design could comprise two pairs of common-source devices fed by the differential output of the mixer, with the drains tied together. The net output current would consist primarily of the second-order term, which could be fed back into the tail current of the input differential pairs. By controlling the amount of second-order current that is fed back, a controllable second-order term can be synthesized.

The analysis in this section does not consider the nonlinearity of the mixer. Since the IF contributed by several harmonics couple back into the mixer, the nonlinearity of the tail mixer is not modified in a manner similar to that of the current devices. Further, this analysis assumes broadband nonlinear mechanisms. Thus, phase shifts given by various nonlinear terms are not considered, and these will limit the degree of cancellation that is possible. If the baseband LPF and the HPF corner frequencies are much higher than the highest IF of interest and sufficiently lower than the RF, the impact of the phase shift is minimized.

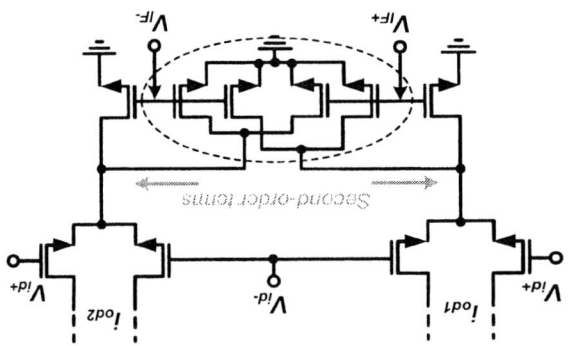

Fig. 5. A simple design modification using two pairs of common-source devices.

IV. SIMULATION RESULTS

The obtained derivation has been verified by simulation using Cadence SpectreRF in a UMC 0.13-μm RF CMOS process with a nominal supply voltage of 1.2 V. The two-tone linearity test was performed for two different models: the receiver down-converter with an ideal mixer and the receiver down-converter with a folded Gilbert cell switching mixer. The IM3 product at 2 MHz was evaluated with two tones of 905 and 908 MHz and an LO signal of 900 MHz.

Using ideal polynomial voltage controlled voltage and current sources to model the long-channel devices of Fig. 1, and an ideal multiplier for the mixer, Eq. (12) was verified exactly, including perfect cancellation of the third order nonlinearity.

To verify this cancellation using device models, we first considered the third-order intermodulation distortion only given by input stages. An ideal mixer based on a polynomial voltage controlled voltage source multiplier was employed in this simulation. Through design of the tail current, the amount of injection of the second-order term (β_2) was suitably controlled. Fig. 6a shows the simulated first- and third-order terms at the output by sweeping β_2 within expected range obtained by hand calculation. As can be observed in Fig. 6a, the minimum IM3 product was achieved when $\beta_2 \approx 105$ m; the

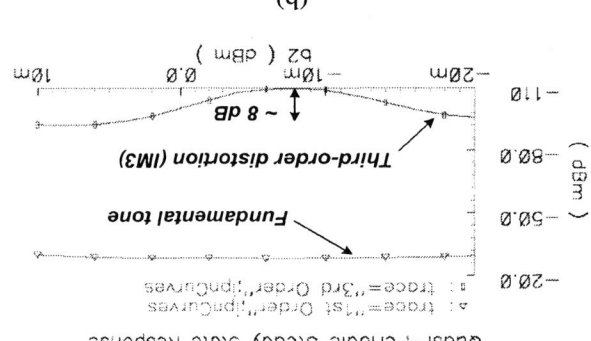

(a)

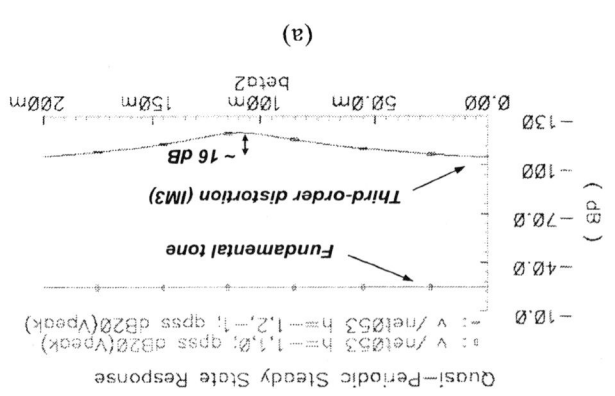

(b)

Fig. 6. First- and third-order terms a versus a second-order nonlinear coefficient of a down-converter (a) with an ideal mixer and (b) with a folded Gilbert cell switching mixer

Fundamental tone was almost invariant. Thus, approximately 8 dB improvement in IIP3 was observed in the simulation.

Two-tone linearity test for cascaded and feedback-based down-converters with a real mixer is shown in Fig. 6b and Fig. 7. For an equivalent comparison, the second amplification stage of both the designs provides identical nonlinear characteristics and a simulated overall conversion gain of 52 dB was achieved from both designs. From Fig. 7, the simulated IIP3s of the cascaded and feedback-based down-converters are −39 dBm and −43 dBm for the peak gain, respectively. Due to cancellation mechanism provided by the second-order nonlinearity of tail current sources, the IIP3 was improved by approximately 4 dB. We believe the third-order intermodulation distortion arising from the mixer switching devices limits the amount of a rejection ratio.

V. CONCLUSION

The third-order nonlinear characteristics of a demonstrated receiver down-converter with multiband feedback and current reuse has been analyzed and verified through circuit simulation. Design modification for further improving the linearity performance of the down-converter has also been discussed.

REFERENCES

[1] J. Han and R. Gharpurey, "A 2.5 mW 900 MHz receiver employing multiband feedback with bias current reuse," IEEE RFIC Symp., Hawaii, June 2007.

[2] P. Sivonen, A. Vilander and A. Parssinen "Cancellation of second-order intermodulation distortion and enhancement of IIP2 in common-source and common-emitter RF transconductors," IEEE Trans. on Circuit and Systems-Part I, vol. 52, no. 2, Feb. 2005

[3] P. R. Gray, P. J. Hurst, S. H. Lewis, and R. G. Meyer, Analysis and Design of Analog Integrated Circuits, 4th ed., John Wiley and Sons, Inc. New York, 2001

[4] T. W. Kim, B. Kim and K. Lee, "Highly linear receiver front-end adopting MOSFET transconductance linearization by multiple gated transistors," IEEE J. Solid-State Circuits, vol. 39, no. 1, pp. 223-229, Jan. 2004

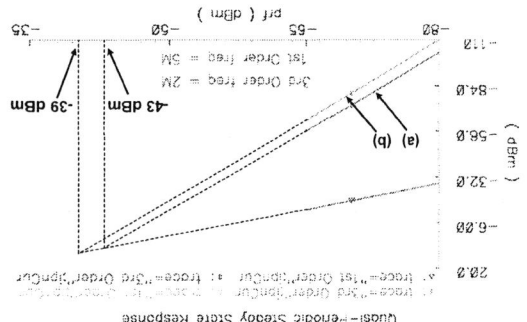

Fig. 7. Simulated IIP3 for (a) cascaded down-converter and (b) feedback-based down-converter

Enhancement of Coexistence Performance in a DRP Based Multi-Radio Environment of a Mobile Phone

Yossi Tsfati, Oren Eliezer, Ran Katz, Yaniv Tzoreff, and Ofer Friedman

Texas Instruments Inc., Wireless Terminals Business Unit

Abstract — The incorporation of Bluetooth and Wireless LAN transceivers in a mobile device poses design challenges associated with their coexistence with the cellular transceiver, as they often need to operate concurrently and their close proximity is inevitable. We present recently developed techniques that transform the RF and analog circuit design complexities, typically encountered when targeting coexistence performance improvements, into the digital domain, and demonstrate how these techniques allow the concurrent operation of multiple transceivers without one adversely impacting the performance of the other. The ideas presented have been implemented in Bluetooth/WLAN transceivers offered by Texas Instruments.

Index Terms – Coexistence, Bluetooth, WLAN, DRP

I. INTRODUCTION

Wireless connectivity solutions, such as Bluetooth and Wireless LAN links, are becoming increasingly prevalent in cellular handsets. The natural drive for lower cost, combined with the integration trend, create challenges of accommodating several co-located connectivity solutions operating simultaneously in very close proximity within the mobile handset. The densely populated radio environment poses additional interferences from devices that are not co-located. In this paper we present techniques that address these coexistence challenges ranging from the physical layer to system level solutions. The techniques, some of which were presented in [1], leverage key benefits of TI's Digital RF Processor (DRP™) technology presented in [2] and [3].

The organization of this paper is as follows: Section II presents the coexistence challenges that are addressed by the presented performance enhancement techniques. The fundamentals of the DRP technology, which are salient to these techniques are discussed in Section III. Section IV describes how the DRP based techniques are applied in order to meet the coexistence challenges.

II. THE COEXISTENCE PROBLEM

Mobile devices often include multiple radios that are required to operate simultaneously and are located in close proximity (on the same PCB or even the same die), resulting in potential mutual interference. Additionally, there could be coexisting transceivers located within range but in different platforms, such that cross-platform collaboration is inhibited. In the analysis of an interference mechanism

experienced in a particular coexistence scenario, we distinguish between the aggressor, which is usually a transceiver's transmitter, and the victim, which is usually an independent coexisting receiver. As most wireless standards today employ TDD channel access methods, each transceiver may serve as both an aggressor and a victim at different instances. This scenario is illustrated in Figure 1, where two transceivers, denoted "A" and "B", are co-located, and the third transmitter, denoted "C" is located in another platform (separated some 1 meter away). The most problematic interference mechanisms, which are the focus of this paper, are when transmitter A interferes with the reception of receiver B and vice-versa. Previous work, such as in [4], identifies known radio standards that potentially interfere with each other when in close proximity or co-located. In [1] several interference mechanisms are discussed and analyzed. To ensure adequate coexistence performance, each one of the transceivers should be carefully designed so as to minimize its potential interference as an aggressor as well as maximize its immunity to interference as a victim. We distinguish between in-band coexistence scenarios, where both transceivers share the same frequency band (for example, an ISM band), and out-of-band scenarios, where the transceivers use non-overlapping frequency bands (for example, an ISM band and a cellular band). Previous work, such as in [4], presents mitigation techniques to be used at the overall product level (PCB, layout and module separation methodologies, etc.) derived from the targeted transceiver performance, whereas the approach presented here focuses on techniques implemented at the SoC level.

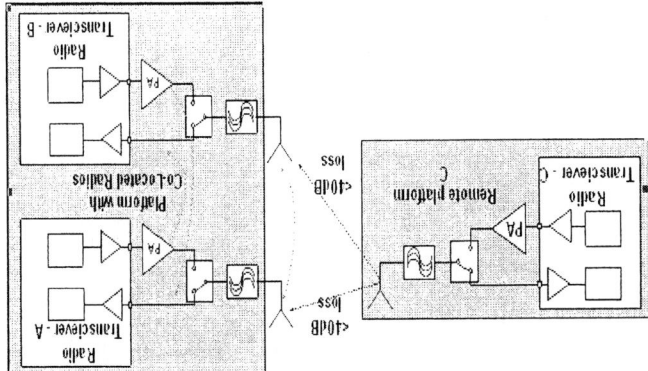

Figure 1 - Mutual interference paths for co-located coexisting transceivers

III. DRP FUNDEMENTALS

This section briefly describes some fundamentals of the DRP technology, focusing on those aspects which are relevant to the coexistence discussion that follows.

The DRP technology was first introduced in 2001, as it was used by Texas Instruments in its first Bluetooth single-chip solution in 130mm CMOS. It is currently also used in TI's GSM system-on-chip (SoC) offering in 90nm CMOS.

A. All Digital PLL

A DRP based polar transmitter is illustrated in the TX block of Figure 2 comprising a DCO, TDC and high speed phase-detection logic. The transmitter is built around an All-Digital PLL (ADPLL), which controls a digitally controlled oscillator (DCO) [2]-[3]. Contrary to conventional implementations [5], the ADPLL is fully digital and has no multiplication operation performed in its phase detector. Consequently, the RF outcome of the ADPLL is of very low spurious content.

B. DRP Transmitter

The transmitter architecture in Figure 2 is based on an R-θ (polar) structure wherein separate paths are dedicated to the amplitude R and phase θ modulations, in contrast with conventional quadrature up-conversion structures. The loop bandwidth of the ADPLL is selected to optimize the phase-noise performance, while a two point modulation structure is employed to accomplish wide modulation bandwidth, which greatly exceeds the loop bandwidth [2]. The amplitude is modulated using a Digital Power Amplifier (DPA) described in [2], [3] and [6]. The DPA circuit, which acts as a digital-to-RF-amplitude converter, is driven by the square wave output of the DCO frequency divider. The amplitude-modulation resolution is further enhanced by employing a high-rate Σ-Δ dithering circuit. Thus, the polar DRP transmitter produces a modulated signal without using conventional mixers.

C. DRP Receiver

The DRP receiver, illustrated in the "RX" block of Figure 2, performs the most critical receiver functions, such as channel selectivity, image rejection and demodulation, in the digital domain. It is built around a high dynamic range oversampling Σ-Δ ADC. The high dynamic range reduces the overall gain needed prior to the digital converter, resulting in better front-end linearity performance and, consequently, enhanced immunity to interfering blockers. RF sampling and discrete-time processing are employed in the front-end prior to the digitization of the signal [2].

IV. LEVERAGING THE DRP CAPABILITIES TO MITIGATE INTERFERENCE EFFECTS

In this section we present methods for addressing some interference mechanisms, while distinguishing between transmitter based techniques, receiver based techniques and system level techniques.

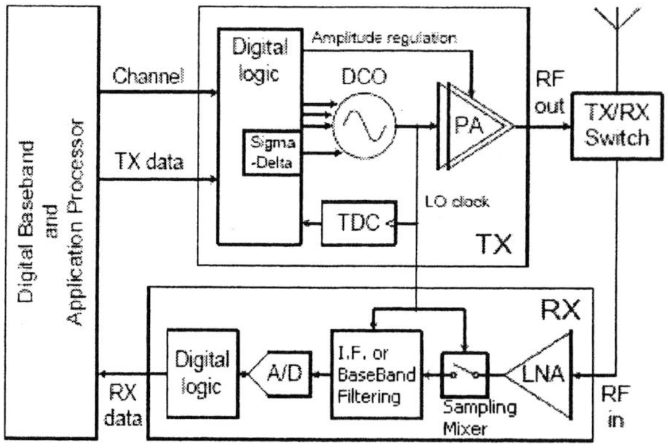

A. Transmitter Techniques

Transmitter techniques are aimed at lowering the impact of the aggressor on the potential victims. In general, a potential aggressor would ideally eliminate any spurious emissions outside of its band of operation and, whenever possible, minimize its emitted power within it. There are various mechanisms that cause spurious emissions, with the dominant ones being generated as a byproduct of the carrier frequency synthesis and modulation, wideband noise originating from the local oscillator's phase noise as well as the modulation mixers and power amplifier, and transient noise caused by the ramp-up and ramp-down of the power amplifier at the beginning and end of a transmission.

1. Low Level of Natural Noise and Spurs

The DRP transmitter benefits from the absence of analog multiplication functions in its modulation path and frequency synthesis circuitry. The primary potential spur creation source is the Σ-Δ clock feed-through into the DCO, frequency modulating it. However, the modulation index for this signal is extremely low since the frequency deviation is measured in kHz and the modulation rate is in the range of hundreds of MHz, making the spur level insignificant (usually <-90dBc) due to the Bessel function decay. Additional potential sources for spurs are clock signals used to drive the high speed digital DRP logic. These are to be designed such that their frequency does not interfere with any of the potential victim frequency bands.

The wideband noise emission is kept low due to the transmitter's use of an R-θ modulation scheme. Due to the absence of mixers in the modulation path, typically being the main contributors to the transmitter's overall noise figure, and noise-pickup, a low noise floor is achieved at the transmitter's output. Contrary to typical levels of out-of-band noise in excess of -135dBc/Hz, the digital power amplifier (DPA), achieves much lower noise floors below -145dBc/Hz, which results in less out-of-band aggressor energy, allowing a Bluetooth transmitter to not desensitize a GSM receiver in the same mobile device. As shown in [1],

Figure 2 - The DRP architecture for a Bluetooth transceiver – a digital polar transmitter and a discrete time receiver

such noise level, when coupled into a victim GSM receiver, causes negligible desensitization.

II. Spectral Shaping of Quantization Noise

The DPA based amplitude modulation path creates wideband quantization noise when modulating a non-constant envelope signal (e.g. EDGE and Bluetooth-EDR signals) since its resolution is finite, subject to the smallest size of the switched drivers that construct the DPA. To minimize the energy of this emission source in specific bands, the transmitter may digitally shape the output noise of the Σ-Δ modulator as suggested in [2], [3] and [7], having an implementation advantage over conventional Σ-Δ architectures [6].

III. Digital Shaping of Ramping Function

It is desirable to minimize the out-of-band transient noise created during the power ramping of the transmitter at the beginning and end of a transmitted packet. For a given limited ramping duration dictated by the protocol, there exists an optimal ramping waveform that would minimize the emissions in the band of interest, which can be implemented by the DRP based transmitter, as shown in [1].

B. Receiver Techniques

Receiver techniques target the minimization of the impact of an existing aggressor on the reception performance of the victim. The aggressor's energy can be divided into two categories: the fundamental carrier that is out of the victim's band, and spurious emissions that fall within the victim's band. The former may cause compression effects that reduce the effective gain and introduce excessive noise and spurs through intermodulation effects. Additional forms of interference are described in [1]. The receiver's linearity and the RF filtering dictate the immunity to the out-of-band interference energy, while the receiver selectivity may affect the performance in the presence of in-band interference energy.

When a strong out-of-band blocker exceeds the front-end's input saturation level, the receiver experiences gain loss resulting in desensitization. The likelihood and severity of desensitization increase as the front-end gain increases.

The impact of gain loss and noise caused by the aggressor's out-of-band energy is mitigated by minimizing the gain in the receiver chain. The DRP receiver uses a high dynamic-range ADC, which allows the use of relatively low analog gain in the receiver's front-end. This provides high robustness to strong out-of-band aggressors and enhanced linearity performance in the presence of close-in interfering signals, since the impact of interference is directly related to the receiver's dynamic range or linearity. The relationship between the receiver gain and its linearity performance is illustrated in [1].

In a DRP based WLAN receiver the extended dynamic range allows the LNA gain to be limited to 26dB, which could accommodate an EVM better than 50dB in the presence of strong in-band and out-of-band blockers.

In contrast, traditional implementations typically require a front-end gain higher than 30dB, resulting in higher susceptibility to blocking.

The wide dynamic range, combined with the high sampling rate of the ADC, enable the receiver to use very simple baseband filtering, such as a simple RC filter, instead of a more complicated LPF that may be necessary when the ADC sampling rate and dynamic range are low. Adjacent channel interference can then be filtered in the digital domain more accurately and efficiently. In addition to the simplification and improvement of the adjacent channel filtering, the higher dynamic-range and sampling rates of the ADC enable the receiver to accommodate wideband observations, which is instrumental for rapid interference detection and avoidance. This may be exploited by the DRP receiver to estimate the parameters of interfering signals in a band of interest. This is then used to optimally set the receiver parameters such that the performance degradation is minimized. One specific setting can be the gain of the receiver back-end. When strong interference is detected, this gain may be reduced, thereby extending the receiver's linearity and minimizing excessive noise that could be created due to compression and intermodulation effects.

The selectivity and rejection of interference may be enhanced through the narrowing of the receiver's bandwidth. However, this could adversely impact the receiver's performance due to the linear distortion the received signal may suffer and the consequent inter-symbol interference (ISI). This often necessitates equalization, which is performed in the digital domain. In general, the filter's rejection is preferably improved by adding more poles, as the increase in the baseband filtering capability is proportional to the number of poles in the filter. Typical implementations of baseband filters with high rejection may be Chebyshev or Butterworth topologies. With these topologies, the number of poles is related with the number of active amplifiers. As the number of poles increases, the number of required amplifiers increases and the current consumption of the filter is increased. This is illustrated for a WLAN receiver chain in Figure 3. As the rejection offered by the analog filter is higher, the ACI performance is improved, whereas the current consumption of the filter is slightly increased.

C. System Level Techniques

System level techniques target the adjustment of system behavior such that it will coexist optimally based on timing and/or frequency information. Time-division coordination between radios is often used, in particular for radios that share the same antenna. For devices that are co-located in the same platform, but do not share an antenna, high-level coordination between the radios may be replaced with independent interference detection, to ensure that a potential aggressor does not transmit during an instance when its transmission or byproducts of it may adversely impact the victim receiver. A primary application of this technique is spectral scanning of the receiver's input and tuning of the victim receiver or aggressing transmitter to avoid collisions.

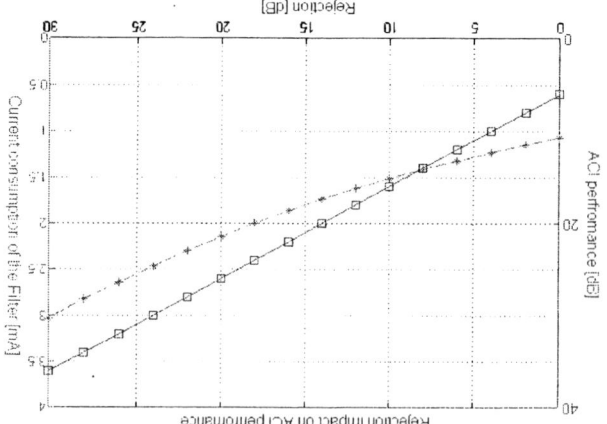

Figure 3 - Effect of rejection of baseband filter on the ACI ratio performance (with squares) and on current consumption (with '*').

Timing-coordination between two different radios that share a single antenna may be employed in a Bluetooth device co-located with a WLAN transceiver, both operating in the 2.4GHz band. This is implemented by controlling the antenna switch in accordance with the active transmitter/receiver, as dictated by a higher-level traffic-prioritizing processor. This time-division mechanism ensures the quality of critical links, such as voice, for both the Bluetooth and WLAN radios by scheduling the WLAN transactions such that the Bluetooth voice quality is not degraded and no packet loss is experienced in the WLAN system. This is accomplished by monitoring WLAN and Bluetooth activity, determining access priority for each device, predicting Bluetooth high priority patterns when they exist, allocating the bandwidth for the appropriate device according to the above, and when needed, reconfiguring the hardware accordingly (turning PA on/off and switching the antenna to the proper position). The BW allocation of the WLAN is done by forcing the timing of the transmissions of the WLAN access-point to inactive Bluetooth periods, through the use of the appropriate WLAN protocol commands. The calculations involved with the multiplexing operation are realized within the embedded processor of the DRP, which primarily serves for control and calibration purposes.

High speed scanning of the in-band frequency spectrum is accomplished in the DRP receiver by leveraging the ADPLL's two point modulation to adjust the receiver LO away from its original settings, without even requiring another frequency acquisition cycle. Using this technique, the ADPLL can perform a frequency step in less than 1 μsec by activating a set of a few varactors in the overall LC tank of the DCO. This high speed scanning ability of the ADPLL is exploited to detect the presence of in-band interference within a minimal duration. Once interference is detected at a particular frequency, an indication is sent to the DRP and MAC processors. The DRP processor uses this information to set the parameters of the receiver chain to optimize the performance (e.g., the receiver gain is adjusted). The MAC processor can use the information to take a proactive action to ensure that the transceiver won't be set to this particular frequency, by using the Adaptive Frequency Hopping (AFH) algorithm that is part of the current Bluetooth specifications. Using the high speed frequency scanning ability of the DRP, the MAC can immediately detect the presence of such in-band interference and quickly remove the interfered frequencies from the hopping sequence used for the Bluetooth communications. This mechanism also serves to enhance the coexistence performance in the presence of a cellular transmission in the 850MHz band having a third harmonic that falls within the 2.4GHz band.

V. CONCLUSION

The recently developed techniques presented here effectively address the coexistence issues that are rapidly emerging in collocated transceivers in mobile handset devices. The combination of the ADPLL, DCO and DPA provides spectral purity for the DRP based transmitter, thus minimizing its interference to other victim receivers operating either in the same band (e.g. Bluetooth and WLAN), or in a distant band.

The high dynamic range of the ADC in the receiver allows the use of low receiver-front-end gain, thus achieving extended linearity and enhanced interference immunity.

Accurate time-division coordination between the Bluetooth and WLAN transceivers is achieved by using the strong computational capabilities of the processor originally embedded in the DRP for control and calibration routines. The ADPLL's ultra-fast frequency tuning capability is exploited to scan the in-band frequency range, detect interference and enhance the coexistence performance in its presence.

REFERENCES

[1] O. Friedman, Y. Tsfati, R. Katz, N. Tal, O. Eliezer "Coexistence Performance Enhancement Techniques for DRP Based WLAN/WPAN and Cellular Radios Collocated in a Mobile Device", CCNSI 2006, Oct. 2006.

[2] R. B. Staszewski, K. Muhammad, D Leipold, et al., "All-digital TX frequency synthesizer and discrete-time receiver for Bluetooth radio in 130-nm CMOS", IEEE Journal of Solid-State Circuits, vol. 39, iss.12, pp. 2278-2291, Dec. 2004.

[3] R. B. Staszewski, K. Muhammad, D Leipold, "Digital RF Processor (DRPrm) for Cellular Radios", NORCHIP Conference 2005, 23rd, 21-22 Nov 2005. Pages 1-4.

[4] M. Gaynor, D. Mathews, "System-in-Package for WLAN/PAN Aids Coexistence with Digital Cellular", High Frequency Electronics, Summit Technical Media, LLC, January, 2005.

[5] N. Filiol, N. Birkett, J. Cherry et al., "A 22mW Bluetooth RF transceiver with direct RF modulation and on-chip IF filtering", Proc of IEEE Solid-State Circuits Conf, pp 202,203, 447, Feb 2001

[6] P. Cruise, C.-M. Hung, R.B. Staszewski, O. Eliezer, S. Rezeq, D. Leipold and K. Maggio, "A digital-to-RF Amplitude converter for GSM/GPRS/EDGE in 90-nm digital CMOS", Proc. Of 2005 IEEE Radio Frequency Integrated Circuits (RFIC) Symp, sec. RMO1A-4, pp. 21-24, June 2005, Long Beach, CA.

[7] V.K. Parikh, G. Feygin at al., "Implementation of high-speed digital band-pass sigma-delta modulator for a wireless transmitter", IEEE Transactions on Circuit and Systems-III: Analog and Digital Signal Processing, vol. 41, No.2, pp. 207-210, March 2005.

WAVEFORM ANALYSIS AND DELAY PREDICTION IN SIMULTANEOUSLY SWITCHING CMOS GATE DRIVEN INDUCTIVELY AND CAPACITIVELY COUPLED ON-CHIP INTERCONNECTS

B.K.Kaushik, S.Sarkar*, R.P.Agarwal* and R.C.Joshi
Indian Institute of Technology-Roorkee, *MITS-Sikar, INDIA
[bkk10dec, rajanfec, joshifcc]@iitr.ernet.in, *sankarsarkar@gmail.com

Abstract: This paper focuses on waveform analysis and delay estimation of a CMOS gate driven capacitively and inductively coupled interconnect for simultaneously switching inputs. A transmission line based coupled model of interconnect is used for analysis. Delays at far-end of victim are estimated for the conditions when the inputs to two coupled interconnects are switching in-phase and out-of-phase. Alpha Power Law model of MOS-transistor is used to represent the transistors in CMOS-driver. The comparison of analytically obtained results with SPICE simulations show that the proposed model captures 90% propagation delay; transition time delay and waveform shape with good accuracy.

1. INTRODUCTION

Drivers have significant impact on crosstalk noise analysis in a Driver-Interconnect-Load (DIL) circuit. In past researchers had modeled crosstalk noise in distributed coupled interconnects by representing the driver CMOS gate by a simple resistor. Similar to other authors, Agarwal *et al.* [1] also assumed the driver impedance to be a linear resistance. Modeling non-linear CMOS driver by an equivalent linear resistive driver leads to discrepancy in results. This can be understood by noting that a transistor in a CMOS gate operates partially in the linear region and partially in the saturation region during switching. It is only in the linear region that a transistor can be accurately approximated by a resistor, where as in the saturation region, the transistor is more accurately modeled as a current source with a parallel high resistance. Furthermore, they [1] considered unipolar input for in-phase switching but incorrectly assumed bipolar inputs (0 to V_{DD} and - V_{DD} to 0) for out-of-phase switching. In fact, in a realistic scenario for CMOS technology, only unipolar inputs (0 to V_{DD} and V_{DD} to 0) are considered. As an improvement over previous crosstalk noise models, we propose a composite DIL model that includes an accurate Alpha power law transistor model [2] with unipolar inputs.

2. DRIVER-INTERCONNECT-LOAD MODEL

The aggressor and victim line CMOS driver transistor can be represented by Alpha Power Law model [2] which is given by

$$I_D = \begin{cases} 0 & V_{GS} \le V_{TO} & : \text{cut-off region} \\ k_l (V_{GS}-V_{TO})^{\alpha/2} V_{DS} & V_{DS} < V_{D-SAT} & : \text{linear region} \\ k_s (V_{GS}-V_{TO})^{\alpha} & V_{DS} \ge V_{D-SAT} & \text{saturation region} \end{cases} \qquad (1)$$

where V_{D-SAT} is drain saturation voltage; k_l and k_s are transconductance parameters in linear and saturation regions respectively; α is velocity saturation index; and V_{TO} is zero bias threshold voltage. The proposed DIL model comprises of the coupled transmission line model [1], represented as

$$V_1 = \left(A_1 e^{-\gamma_e z} + A_2 e^{\gamma_e z}\right) + \left(A_3 e^{-\gamma_o z} + A_4 e^{\gamma_o z}\right)$$

$$V_2 = \left(A_1 e^{-\gamma_e z} + A_2 e^{\gamma_e z}\right) - \left(A_3 e^{-\gamma_o z} + A_4 e^{\gamma_o z}\right)$$

$$I_1 = (1/Z_{0e})\left(A_1 e^{-\gamma_e z} - A_2 e^{\gamma_e z}\right) + (1/Z_{0o})\left(A_3 e^{-\gamma_o z} - A_4 e^{\gamma_o z}\right)$$

$$I_2 = (1/Z_{0e})\left(A_1 e^{-\gamma_e z} - A_2 e^{\gamma_e z}\right) - (1/Z_{0o})\left(A_3 e^{-\gamma_o z} - A_4 e^{\gamma_o z}\right) \qquad (2)$$

Here, $V_1(z,t)$, $I_1(z,t)$, and $V_2(z,t)$, $I_2(z,t)$ are voltage and current waveforms on lines 1 (aggressor) and 2 (victim) respectively. The A_i's are constants whose values are obtained from the boundary conditions. The constants γ_e, Z_{0e}, and γ_o, Z_{0o} are propagation constant and characteristic impedance for even and odd modes. The coupled interconnects are assumed to be symmetric. Receivers at the far-end of the lines are modeled as lumped capacitive loads.

An input signal to CMOS driver is categorized to fast or slow ramp depending on the state PMOS (for falling ramp)/NMOS (for rising ramp) device attains when the input voltage reaches its final value. If the MOSFET continues to operate in saturation region when the input ramp has reached its final value, the ramp is called fast. On the other hand, if the MOSFET switches to linear region of operation before the input ramp attains its final value, the input ramp is said to be slow. The slope of input ramp is $M_s = V_{DD}/\tau$.

3. IN-PHASE SWITCHING

Under in-phase switching, we consider a case where the inputs to aggressor and victim lines are falling (Figure 1). In order to obtain analytically the far-end voltage on victim line, the CMOS has to pass through four regions of operation in each case of fast and slow input ramp.

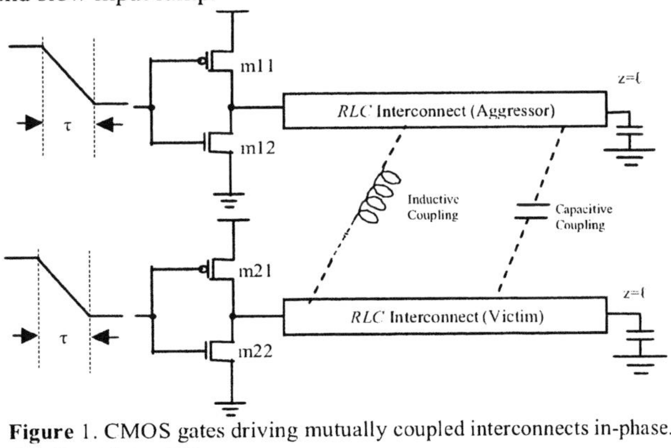

Figure 1. CMOS gates driving mutually coupled interconnects in-phase.

3.1 Fast Input Ramp

Region-I ($0 < t < t_1$): During this interval of time, m11 and m21 are in cutoff state; m12 and m22 operates in linear region. Since the voltage drop across drain and source is very low, the current through m12 and m22 is negligible. Thus, NMOS of aggressor and victim drivers are in high impedance state. The instant of time t_1, when PMOS just enters saturation region, can be obtained directly by knowing the threshold voltage of the device. The collective gate to drain capacitance of NMOS and PMOS (C_m) is negligible in comparison to the interconnect line capacitance. Thus, during the short transition period of input the interconnect line capacitance is negligibly charged/discharged through C_m. Using these boundary conditions and through the analysis of [1]

$$(A_1/Z_{0o}) + (A_3/Z_{0o}) = 0 \; ; \text{ and } \; (A_1/Z_{0o}) - (A_3/Z_{0o}) = 0 \quad (3)$$

Solving the set of equations (3), we obtain $A_1 = 0$; and $A_3 = 0$.

Region-II ($t_1 < \tau < t_2$): During this region, m11 and m21 operates in saturation region. Although the voltage at node $z = 0$ rises during this stage, but does not rise appreciably to drive significant current through NMOS. The current through the PMOS parasitic gate-to-drain capacitance is considered to be negligible as compared to the drain current, I_{Dsp} of the PMOS.

Therefore, $I_{Dsp} \equiv I_1(z = 0)$. For PMOS of the driver

$$I_{Dsp}(sat) = k_s(V_{Gs} - V_{TO})^\alpha \quad (4)$$

where $\hat{k}_s = I_{Do}/(V_{DD} - V_{TO})^\alpha$ and $V_{Gs} = (V_{DD} - V_{TO})/\tau \cdot t = M_s t$ (5)

From (4) and [1], $(A_1/Z_{0o}) + (A_3/Z_{0o}) = k_{s1}(M_s t - V_{TO1})^{\alpha 1}$ (6)

$(A_1/Z_{0o}) - (A_3/Z_{0o}) = k_{s2}(M_s t - V_{TO2})^{\alpha 2}$ (7)

k_{s1} and V_{TO1} corresponds to PMOS m11, whereas k_{s2} and V_{TO2} corresponds to PMOS m21. Solving equations (6) and (7)

$$A_1 = (1/2)\cdot Z_{0o}\left[k_{s1}(M_s t - V_{TO1})^{\alpha 1} + k_{s2}(M_s t - V_{TO2})^{\alpha 2}\right] \quad (8)$$

$$A_3 = (1/2)\cdot Z_{0o}\left[k_{s1}(M_s t - V_{TO1})^{\alpha 1} - k_{s2}(M_s t - V_{TO2})^{\alpha 2}\right] \quad (9)$$

For homogeneous drivers ($k_{s1} = k_{s2} = k_s$; $V_{TO1} = V_{TO2} = V_{TO}$ and $\alpha 1 = \alpha 2 = \alpha$), $A_1 = Z_{0o}k_s(M_s t - V_{TO})^\alpha$ and $A_3 = 0$.

Region-III ($\tau < t < t_2$): Over this region, PMOS m11 and m21 are in saturation and the input ramp has reached its final value. At time t_2, PMOS enters into linear region of operation. Since the input has reached its final value the gate-to-source parasitic capacitances are d.c. biased, and therefore negligible current flows through them. The currents $I_{Ds,p}$ and $I_1(z = 0)$ at the near end of the interconnect NMOS (m12/m22) is cut-off and therefore no current flows through it.

If, t_2 is the instant when $|V_i(z = 0)| = |V_{DD}|$; and $I_{Ds,p}(z = 0) = I_{Ds,p}(sat)$ then by (1) and because the input has reached its final value $V_{DS}(sat) = (k_s/k_l)\cdot(V_{DD} - V_{TO})^{\alpha/2}$ and $I_{Ds,p}(sat) = I_{Do}$. Therefore at time t_2,

$V_i(z = 0) = V_{DD} - (k_s/k_l)\cdot(V_{DD} - V_{TO})^{\alpha/2}$ and $I_1(z = 0) = I_{Do}$.

With boundary conditions of this region and [1],

$(A_1/Z_{0o}) + (A_3/Z_{0o}) = I_{Do1} = I_{Do}$; and $(A_1/Z_{0o}) - (A_3/Z_{0o}) = I_{Do2}$ (10)

I_{Do1} and I_{Do2} are saturation region current for m11 and m21 respectively. Solving (10), we obtain

$A_1 = (1/2)\cdot Z_{0o}(I_{Do1} + I_{Do2})$; and $A_3 = (1/2)\cdot Z_{0o}(I_{Do1} - I_{Do2})$ (11)

For homogeneous drivers ($I_{Do1} = I_{Do2} = I_{Do}$), the equation (11) reduces to $A_1 = Z_{0o}I_{Do}$; and $A_3 = 0$.

Region-IV ($t > t_2$): This region of input transition time has the PMOS operation in its linear region, whereas, the NMOS remains in cut-off state. The near ends of aggressor and victim lines at node z=0 are directly connected to V_{DD} through linear resistance R_s and R_r respectively. For this part of the input linear transition time,

$$A_1 = \frac{(R_r + 2Z_{0o} + R_s)V_{DD}Z_{0c}}{2R_rR_s + R_rZ_{0c} + R_sZ_{0o} + 2Z_{0c}Z_{0o} + R_rZ_{0o} + R_sZ_{0c}} \quad (12)$$

$$A_3 = \frac{V_{DD}Z_{0o}(R_s - R_r)}{2R_rR_s + R_rZ_{0c} + R_sZ_{0o} + 2Z_{0c}Z_{0o} + R_rZ_{0o} + R_sZ_{0c}} \quad (13)$$

For homogeneous drivers $R_s = R_r$, which reduces (12) and (13) to

$A_1 = V_{DD}\cdot Z_{0o}/(R_s + Z_{0c})$ and $A_3 = 0$.

3.2 Slow Input Ramp

Region-I ($0 < t < t_1$): The operating condition of the structure in region-1 is same as for fast input ramps.

Region-II ($t_1 < \tau < t_2$): The operating condition of this region is also similar to that of fast inputs; however, it extends from time t_1 to t_2, where $t_1 < \tau$, t_2 is the time when $V_1(z = 0) = V_{DD} - V_{DS}(sat)$

where $V_{DS}(sat) = (k_s/k_l)(M_s t - V_{TO})^{\alpha/2}$ (14)

$V_1(z = 0)$ is evaluated with help of A_1 and A_3. With passage of time, the values of $V_1(z = 0)$ and $V_{DS}(sat)$ rises; whereas $|V_{DD} - V_1(z = 0)|$ falls. The magnitude of $|V_{DD} - V_1(z = 0)|$ will be equal to $V_{DS}(sat)$ at time t_2.

Region-III ($t_2 < \tau < t$): During this region, the PMOS transistor operates in linear mode while the input is still a ramp. Using (1) for linear region of operation, the resistance of PMOS is

$$R_{sp}(t) = \left[I_{Do}\cdot(V_{DD} - V_{TO})^{-\alpha/2}/V_{Do}\right](M_s t - V_{TO}) \quad (15)$$

Applying boundary conditions, A_1 and A_3 are obtained as

$$A_1 = \frac{(R_{sp2} + 2Z_{0o} + R_{sp1})V_s Z_{0c}}{2R_{sp2}R_{sp1} + R_{sp1}Z_{0c} + R_{sp2}Z_{0o} + 2Z_{0c}Z_{0o} + R_{sp2}Z_{0o} + R_{sp1}Z_{0c}}$$

$$A_3 = \frac{(R_{sp1} - R_{sp2})V_s Z_{0c}}{2R_{sp2}R_{sp1} + R_{sp1}Z_{0c} + R_{sp2}Z_{0o} + 2Z_{0c}Z_{0o} + R_{sp2}Z_{0o} + R_{sp1}Z_{0c}}$$

For homogeneous drivers, $R_{sp1} = R_{sp2} = R_{sp}$ which results in

$A_1 = V_s\cdot Z_{0c}/(R_{sp} + Z_{0o})$; and $A_3 = 0$.

Region-IV: Region-IV is solved exactly as for fast ramp.

4. OUT-OF-PHASE SWITCHING

Under out-of-phase switching, it is observed that only fast input ramp is valid in majority of cases. This is because of Miller's effect where the coupling capacitance is effectively doubled for out-of-phase switching. Due to this the charging/discharging of interconnect line is quite slow, which inhibits MOS transistor operation in linear region during the transition period of input ramp. Thus, for out-of-phase switching, only fast input ramp case is considered.

Under in-phase switching for falling input signals, the analysis carried out focused on only PMOS transistors through all regions of operation. Since the NMOS transistors were neglected, therefore they can be called as Recessive transistor, whereas PMOS transistor that contributed in major for even (A_1) and odd (A_3) waves can be termed as Dominant transistor. Similarly, for rising input the PMOS and NMOS transistors are called recessive and dominant transistors respectively.

In sharp contrast to in-phase switching, all MOSFETs (either dominant or recessive) contribute towards even and odd waves under out-of-phase switching condition. For rising and falling inputs to aggressor and victim lines respectively, transistors m12 (NMOS) and m21 (PMOS) are dominant, whereas transistors m11 (PMOS) and m22 (NMOS) are recessive. The CMOS drivers of aggressor and victim lines have to pass through ten regions of operation for the fast input ramp case. These regions of operation are shown in Table 1. The inputs to aggressor and victim lines are rising and falling respectively.

Table 1 Generalized regions of operation under out-of-phase switching.

Region	Time		Dominant Transistors		Recessive Transistors	
	From	To	m12-NMOS	m21-PMOS	m11-PMOS	m22-NMOS
I	0	t_1	Cut-off	Cut-off	Linear	Linear
II	t_1	t_2	Saturation	Cut-off	Cut-off	Linear
III	t_2	t_3	Saturation	Linear	Linear	Linear
IV	t_3	t_4	Saturation	Saturation	Saturation	Linear
V	t_4	t_5	Saturation	Saturation	Cut-off	Linear
VI	t_5	t_6	Saturation	Saturation	Cut-off	Saturation
VII	t_6	τ	Saturation	Saturation	Cut-off	Cut-off
VIII	τ	t_7	Saturation	Saturation	Cut-off	Cut-off
IX	t_7	t_8	Saturation	Linear	Cut-off	Cut-off
X	t_8	End	Linear	Linear	Cut-off	Cut-off

In the analysis to follow, contributions towards even (A_1) and odd (A_3) waves through dominant transistors are represented as A_{1d} and A_{3d}, and similar contributions through recessive transistors are A_{1r} and A_{3r}. The contributions of dominant and recessive transistors are added up for finally generating even and odd waves. Every region of operation dealt will show each contribution.

Region-I ($0 < t < t_1$) : Using the boundary conditions [1] it can be shown that $A_{1d}=A_{3d}=0$. The instant of time t_1, when NMOS m12 just enters saturation region, can be obtained by knowing the threshold voltage of the device.

Recessive Transistors: Both transistors are in linear region and the gate-to-source voltages of both transistors are varying with time.

$$A_{1r} = \frac{(R_{22}+Z_0)V_{DD}Z_{0c}}{-2R_{22}R_{11}+R_{22}Z_{0c}+R_{11}Z_{0o}+2Z_0^2Z_{0o}+R_{22}Z_0^2+R_{11}Z_{0c}}$$

$$A_{3r} = \frac{(R_{22}+Z_0)V_{DD}Z_{0o}}{-2R_{22}R_{11}+R_{22}Z_{0c}+R_{11}Z_{0o}+2Z_0^2Z_{0o}+R_{22}Z_0^2+R_{11}Z_{0c}}$$

Here R_{11} and R_{22} corresponds to transistors m11 and m22 respectively. Similarly, the transistor parameters (such as V_{To}, V_{Do}, I_{Do}, k_s, k_l and α) with subscript 11, 12, 21 and 22 correspond to transistor m11, m12, m21 and m22 respectively.

Using (1) the resistances during this region of operation are

$$R_{11} = \frac{V_{Do,11}}{I_{Do,11}(V_{DD}-V_{To,11})}\cdot(V_{DD}-M_s \cdot r - V_{To,11})^{-\alpha_{11}/2}$$

$$R_{22} = \frac{V_{Do,22}}{I_{Do,22}(V_{DD}-V_{To,22})}\cdot(V_{DD}-M_s \cdot r - V_{To,22})^{-\alpha_{22}/2}$$

Region-II ($t_1 < t < t_2$) : Dominant Transistors: Using boundary conditions in [1] and equation (1), we obtain $A_{1d}=A_{3d}=0$. The time

t_2, when m21 enters saturation region, can be obtained by knowing the threshold voltage of the device.

Region-III ($t_2 < t < t_3$) : Dominant Transistors: Using boundary conditions and through the analysis of [1]

$$A_{1d}(t) = (1/2).Z_{0c}.(K_{12}-K_{21}) \; ; \text{ and } A_{3d}(t) = (1/2).Z_{0c}.(K_{12}+K_{21})$$

where $K_{12} = k_{s,12}(M_s \cdot r - V_{To,12})^{\alpha_{12}}$ and $K_{21} = k_{s,21}(M_s \cdot r - V_{To,21})^{\alpha_{21}}$

Recessive Transistors: A_{1r} and A_{3r} are obtained exactly as that in region-I. However the time t_3 is found by following analysis. As per Alpha power law model [2] and (1), the drain to source voltage of m11 during saturation region is given as

$$V_{ds,11}(sat) = (k_{s,11}/k_{l,11})(V_{DD}-M_s \cdot r - V_{To,11})^{\alpha_{11}/2}$$

The time instant t_3 is reached when $V_1(z=0) = V_{DD} - V_{ds,11}(sat)$

Region-IV ($t_3 < t < t_4$) : Dominant Transistors: During this region, A_{1d} and A_{3d} are obtained exactly as that in region-III.

Recessive Transistors: Using boundary conditions

$$A_{1r} = K_{11}Z_{0c}(R_{22}+Z_{0o})/(Z_{0o}+2R_{22}+Z_{0c})$$

$$A_{3r} = K_{11}Z_{0o}(R_{22}+Z_{0o})/(Z_{0o}+2R_{22}+Z_{0c})$$

where $K_{11} = k_{s,11}(V_{DD}-M_s \cdot r - V_{To,11})^{\alpha_{11}}$. Time t_4 when m11 enters cut-off, is obtained by knowing the threshold voltage of device.

Region-V ($t_4 < t < t_5$) : Dominant Transistors: A_{1d} and A_{3d} are obtained exactly as in region-III.

Recessive Transistors: Applying boundary conditions and subsequent solving results in $A_{1r} = 0$; and $A_{3r} = 0$.

The time t_5 is an instant when $V_2(z=0) = V_{ds,22}(sat)$

where, $V_{ds,22}(sat) = (k_{s,22}/k_{l,22})(V_{DD}-M_s \cdot r - V_{To,22})^{\alpha_{22}/2}$

$V_2(z=0)$ is evaluated with help of A_1 and A_3 .

Region-VI ($t_5 < t < t_6$) : Dominant Transistors: During this region, A_{1d} and A_{3d} are obtained exactly as that in region-III.

Recessive Transistors: $A_{1r}=A_{3r}=0$.

Time t_6, when m22 just enters cut-off region, can be obtained directly by knowing the threshold voltage of the device.

Region-VII ($t_6 < t < \tau$) : Dominant Transistors: During this region, A_{1d} and A_{3d} are obtained exactly as that in region-III.

Recessive Transistors: $A_{1r}=A_{3r}=0$.

Region-VIII ($\tau < t < t_7$) : Dominant Transistors:

$$A_{1d} = (Z_{0c}/2).(I_{Do,12}-I_{Do,21}) \; ; \; A_{3d} = (Z_{0c}/2).(I_{Do,12}+I_{Do,21})$$

t_7 is obtained for condition when $V_2(z=0) = V_{DD} - V_{ds,21}(sat)$

where, $V_{ds,21}(sat) = (k_{s,21}/k_{l,21})(M_s \cdot r - V_{To,21})^{\alpha_{21}/2}$

$V_2(z=0)$ is evaluated with help of A_1 and A_3 .

Recessive Transistors: This region is solved as region-VII.

Region-IX ($t_7 < t < t_8$) : Dominant Transistors:

$$A_{1d} = Z_{0c}\left[\frac{R_{21}.I_{Do,12}+Z_{0c}.I_{Do,12}-V_{DD}}{(R_{21}.(Z_{0c}/Z_{0o})+2.(R_{21}/Z_{0o})+1)} - I_{Do,12}\right]$$

$$A_{3d} = \frac{R_{21}.I_{Do,12}+Z_{0c}.I_{Do,12}-V_{DD}}{(R_{21}.(Z_{0c}/Z_{0o})+2.(R_{21}/Z_{0o})+1)} \quad \text{where } R_{21} = \frac{V_{Do,21}}{I_{Do,21}}$$

Time t_8 is found under condition when, $V_1(z=0) = V_{ds,12}(sat)$

where $V_{Ds,12}(sat) = (k_{s,12}/k_{l,12})\left(M_s.1 - V_{To,12}\right)^{\alpha_{12}/2}$

Recessive Transistors: This region is solved as region-VII.

Region-X ($t > t_8$) : Dominant Transistor:

$$A_{1d} = Z_{2o}.(R_{c12} + Z_{2o})/D \; ; \; A_{3d} = -Z_{2o}.(R_{c12} + Z_{2o}).V_{DD}/D$$

where $D = (R_{c21} + Z_{2o}).(R_{c12} + Z_{2o}) + (R_{2o} + Z_{2o}).(R_{21} + Z_{2o})$

and $R_{c12} = V_{Do,12}/I_{Do,12}$

Recessive Transistors: This is solved exactly as region-VII.

5. COMPARISON OF LINEAR AND NON-LINEAR DRIVER OPERATION

This section compares the performance of linear resistive driver model with a non-linear CMOS driver model in a DIL circuit. The coupled interconnects have length, width and spacing of 2mm, 1.2μm and 0.4μm respectively. The parasitics associated with interconnect are L_{tot} =2.15nH, C_{tot} =257fF, C_c =184fF and M=1.68nH. The rise time (τ) of the input is 50ps. For CMOS driver, data of an IBM 0.13μm, 1.2V and Level-49 technology are used. Furthermore, the CMOS driver has PMOS width (W_p) double than NMOS width (W_n). Figures 2 and 3 demonstrates the comparison of the waveforms generated by 1) proposed analytical model using non-linear driver; 2) SPICE simulations using CMOS driver; and 3) the analytical model using equivalent linear region resistance driver for in-phase (W_p=40μm, W_n=20μm) and out-of-phase (W_p=70μm, W_n=35μm) switching respectively. Table 2 shows the transition time delays using either linear driver or non-linear driver. Different PMOS widths are used under in-phase (slow and fast ramps) and out-of-phase (fast ramp) switching. Based on results obtained in Table 2, the maximum and average errors involved are shown in Table 3. The maximum and average errors with linear driver are significantly high. Compared to resistive driver, the waveform shape and delay is more effectively estimated using non-linear driver with respect to (wrt) SPICE simulations.

6. CONCLUSIONS

This paper analyzed waveform at far-end of victim for in-phase and out-of-phase switching with CMOS drivers for unipolar inputs. The α-power law model of MOS-transistor is used to represent a transistor in CMOS-driver. The comparisons of the analytical results with SPICE for CMOS gate driven interconnect shows that the model captures 90% propagation delay; transition time delay and waveform shape quite well. It is concluded that the modeling of CMOS gate driver by a linear resistance, as was done previously by researchers can lead to enormous errors. Our proposed composite DIL model that includes an accurate transistor model for analyzing output noise waveform is much superior.

REFERENCES

[1] K.Agarwal, D.Sylvester, D.Blaauw, "Modeling and analysis of crosstalk noise in coupled RLC interconnects," *IEEE Trans. on CAD of Integrated Circuits and Systems,* vol. 25, pp.892 – 901, 2006.

[2] T. Sakurai and A. R. Newton, "Alpha-power law MOSFET model and its applications to CMOS inverter delay and other formulas," *IEEE J. Solid-State Circuits,* vol. 25, pp. 584-594, 1990.

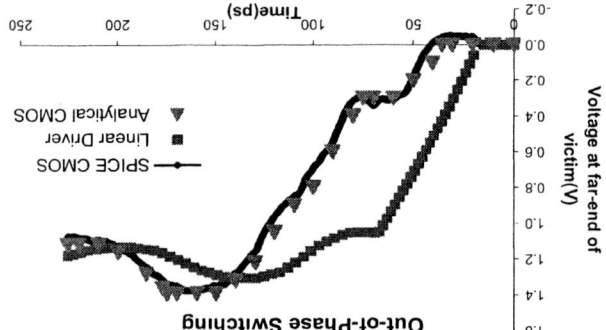

Figure 2 Comparison of waveforms under in-phase switching

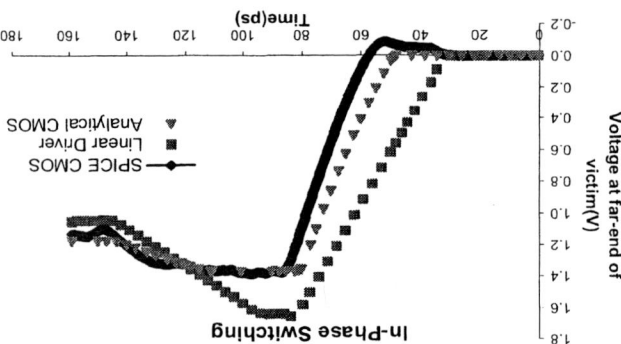

Figure 3 Comparison of waveforms under out-of-phase switching

Table 3 Maximum and Average Errors (wrt SPICE) involved for 90% Propagation Delay (PD), Transition Time Delay (TTD) under in-phase and out-of-phase switching for Linear and Non-Linear Drivers

Switching	Delay	Errors with Non-Linear Driver (%)		Errors with Linear Driver (%)	
		Max.	Avg.	Max.	Avg.
In-Phase	PD	7.86	6.55	20.56	20.4
	TTD	14.4	13.9	93.84	60.35
Out-of-Phase	PD	6.85	4.09	47.88	40.27
	TTD	3.64	3.13	42.4	30.91

Table 2 Computational error involved for 90% propagation time and transition time delays (ps) for resistive and CMOS driver wrt SPICE simulation.

Switching Type	Ramp Type	PMOS Width (nm)	SPICE 90% Prop. Delay	SPICE Trans-ition Delay	Proposed Model 90% Prop. Delay	Proposed Model Transition Delay	Agarwal et al model [1] 90% Prop. Delay	Agarwal et al model [1] Transition Delay	% Error Our Model 90% Prop. Delay	% Error Our Model Transition Delay	% Error 90% Prop. Delay [1]	% Error Transition Delay [1]
Out-of-phase	Fast	100	107.13	68.7	104.93	66.98	55.84	39.57	2.05	2.50	47.88	42.40
	Fast	70	121.23	78.64	118.43	76.36	83.53	66.46	2.31	2.90	31.10	15.49
	Fast	40	162.92	118.87	154.5	114.73	102.03	83.36	3.48	5.17	37.37	37.87
	Fast	20	303.45	225.95	275.8	217.73	167.74	144.93	3.64	6.85	44.72	35.86
In-phase	Fast	35	76.74	20.69	72.47	23.53	61.01	30.61	5.56	-13.73	20.50	-47.95
	Fast	33	77.65	21.51	73.04	24.5	61.7	31.24	5.94	-13.96	20.54	-45.23
	Slow	70	68.78	12.98	63.37	14.79	55.03	25.16	7.86	-13.94	16.99	-93.84
	Slow	40	74.79	18.88	69.65	21.6	59.41	29.15	6.87	-14.41	20.56	-54.40

Design of a Multi-Context FPVLSI based on an Asynchronous Bit-Serial Architecture

Waidyasooriya Hasitha Muthumala, Masanori Hariyama and Michitaka Kameyama

Graduate School of Information Sciences, Tohoku University

Aoba 6-6-05, Aramaki, Aoba, Sendai, Miyagi,980-8579, Japan

Email: {hasitha@kameyama, hariyama@, kameyama@}ecei.tohoku.ac.jp

Abstract—This paper presents a novel asynchronous bit-serial architecture for multi-context field programmable VLSIs (MC-FPVLSI). Conventional MC-FPVLSIs use global wires to distribute the context-ID signal. As a result, hardware utilization ratio decreases, since it is impossible to execute different contexts simultaneously. They also have a high power consumption and high area overhead due to the clock tree and context ID trees. The proposed MC-FPVLSI eliminates the clock tree and global context ID trees completely. It uses a locally distributed context-ID signal and therefore, partial reconfiguration and simultaneous execution of different contexts are possible. It also uses the same wires to transfer the data and context ID signal, so that the area can be reduced further. The proposed architecture is designed using 6-metal 1-poly 90nm CMOS process technology.

Keywords—*dynamically reconfigurable, FPGA, self timing*

I. INTRODUCTION

The dynamically reconfigurable field programmable VLSIs (DR-FPVLSI) provide more cost-effective implementations than the conventional field programmable gate arrays (FPGA). One of the typical DR-FPVLSI architecture is a multi-context FPVLSI (MC-FPVLSI). It has multiple memory bits belong to different configuration planes. The context ID signal selects one of them as the configuration bit. A structure of an MC-FPVLSI is shown in Fig.1. MC-FPVLSIs are said to have a high hardware utilization ratio, because their interconnections are shared between contexts and they can be reconfigured as different circuits in real time. However, since they have only global context-ID wires, the context-ID of the whole chip has to be changed at the same time. As a result, each context has to wait until the execution of its previous context is completed, even though there are adequate and unused hardware resources. Therefore, it is very difficult to achieve a high hardware utilization ratio. The context-ID wires are dedicated wires that connected to every logic block and switch block in the whole chip. Therefore, they need a structure close to a typical clock tree to achieve fast context switching. When the number of context-IDs are increasing, more context-ID wires are needed and it leads to a high area overhead and high power consumption. To solve these problems, we propose a novel MC-FPVLSI based on an asynchronous bit-serial architecture, which can be reconfigured in partially using locally distributed context-ID signals. To the best of the authors knowledge, this is the first ever implementation of an asynchronous MC-FPVLSI architecture.

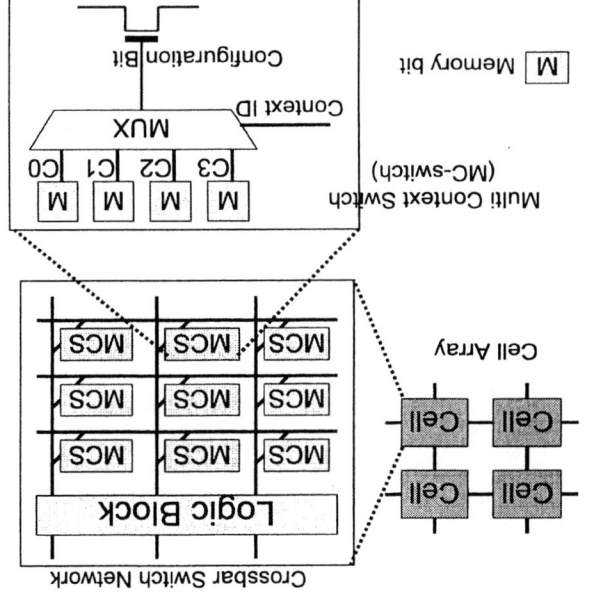

Fig. 1. Structure of a conventional MC-FPVLSI

In this paper we target the following areas in the conventional MC-FPGAs.

- Low hardware utilization ratio due to the global context-IDs that prevents partial reconfiguration.
- High static and dynamic power consumption caused by the clock trees and context-ID trees.
- High area overhead of the context-ID trees when the number of contexts are large.

We propose an asynchronous architecture to eliminate the clock tree completely. Then we remove the global context-ID trees by adopting a bit-serial architecture which can distribute the context-ID signals locally. As a result, context-IDs can be changed partially, so that different contexts can be executed simultaneously in different areas on the chip to increase the hardware utilization ratio. To reduce the area further, we use the same data lines to transfer the context-ID signals. Moreover, the proposed MC-FPVLSI can transfer any number of context-ID-bits without any additional interconnection overhead, unlike the conventional one that requires more context-ID lines when the number of contexts are increased.

II. ARCHITECTURE

A. Data and context-ID encoding scheme

Asynchronous encoding schemes can be classified into:

- **Bundled-data encoding**
- **Delay-insensitive encoding**

Fig.2 shows the overall architecture of the bundled-data encoding. It splits the data and request into separate wires. Data are sent similar to the synchronous architecture, where n-bits of data needs a number n of wires. However, an explicit delay is inserted to the request signal to ensure that the request is received only after the valid data are available. This method needs only $(n + 2)$ wires to send n-bits of data, therefore it can be implemented by using a relatively small area overhead. However, it requires the delay values to operate reliably. If the data path is fixed, it is possible to determine the delay constraint. However, in FPVLSIs, it is not easy to always meet the delay constraint since the data path is programmable.

Therefore we choose delay-insensitive encoding [2], since it does not require any delay insertion. In the delay-insensitive encoding, two wires are used for each data bit as data and request signals, unlike the bundled-data encoding where the request signal is shared between all the data bits. We use level encoded 2-phase dual-rail encoding (LEDR) to encode the data and request signals, since it is very fast data transfer method among the delay-insensitive encoding schemes [3].

Fig.3 shows the LEDR encoded signals (V, R) and their decoded values, data and phase. Fig.4 shows the data transfer based on LEDR encoding. The data bit equals to V and the phase equals to the XOR of V and R $(V \oplus R)$.

Since we want to use the same wires to transfer both data and context-ID signals, we need an another signal to separate the context-ID bits from the data bits. For this purpose, we use a new signal denoted by C as shown in Fig.5. In this paper, we use the term "CVR encoding" for this encoding scheme, since it uses 3 signals C, V, and R to transfer a data bit or a context-ID bit. The encoded signals and their decoded values are shown in Fig.5. The phase equals to $[C \oplus V \oplus R]$ and the data bit equals to V. If the phase is changed, it indicates a new signal is received. If C is "0", the current bit is a data bit, and otherwise it is a context-ID bit.

An example of a bit-serial data and context-ID transfer is shown in Fig.6. At the beginning 3 data bits [0 0 1] are received in a bit-serial manner. Then the value of C changes from "0" to "1" indicating that the next bit is a context-ID-bit. After that, the context-ID-bits are received one after another. In this example, the received context-ID equals to 3 [0,1,1]. After all 3 context-ID-bits are received, the value of C changes from "1" to "0" indicating the next bit is a data bit.

B. Architecture of the logic block

Asynchronous cell based on bundled-data encoding is proposed in [4] for single context FPGAs. However, our proposed architecture is based on delay insensitive encoding and can handle multiple contexts. The block diagram of a logic block (LB) in the proposed MC-FPVLSI is shown in the Fig.7. The LB has 3 main components:

1) Receiver
2) Transmitter
3) Multi-context look-up-table (LUT)

The receiver circuit receives the C,V,R values from previous LBs. When both input C values are "0", receiver passes the V, R values to the LUT. After the C values change from "0" to "1", the context-ID values are stored in the shift register in the receiver. After a context-ID bit is saved in the shift register, receiver sends the acknowledge (ACK) signals to the previous LBs, indicating it is ready to receive another bit. After all the context-ID bits are received, the transmitter starts sending the appropriate V, R values of the context-ID bits to the multiplexers. The multiplexers select the LUT output when C is "0" and the transmitter output when C is "1". The control

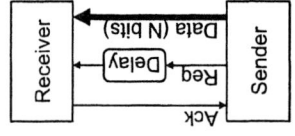

Fig. 2. Bundled-data encoding

Fig. 3. LEDR encoding

Encoded signal		Data	Phase
V	R		
0	0	0	0
1	1	1	0
0	1	0	1
1	0	1	1

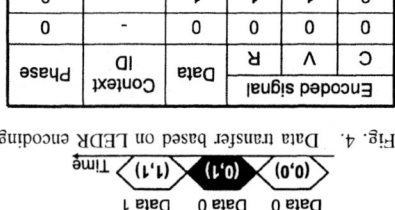

Fig. 4. Data transfer based on LEDR encoding.

Fig. 5. CVR encoding

Encoded signal			Data	Context ID	Phase
C	V	R			
0	0	0	0	-	0
0	1	1	1	-	0
0	0	1	0	-	1
0	1	0	1	-	1
1	0	0	-	0	1
1	1	1	-	1	1
1	0	1	-	0	0
1	1	0	-	1	0

Fig. 6. Data transfer based on CVR encoding

Signal	Data			Context ID				
C	0	0	0	1	1	1	1	0
V	0	0	1	1	0	1	1	1
R	1	0	0	0	0	0	1	1

signal of the multiplexer is delayed for one cycle, so that it can select data and context-ID accurately as shown in Fig.6.

The architecture of a 2-input-8-context LUT is shown in Fig.8. It has 4 memory bits per each context to function as a programmable 2-input logic gate. When the appropriate configuration plane is selected by the context-ID-bits and the multiplexers, the selected memory bits are passed to the decoders. The structure of the decoder-11 in Fig.8 is shown in Fig.9. When the phase of the both inputs ([Va,Ra] and [Vb,Rb]) are "0", both V and R are equal to the memory bit M11, so that the output phase equals to the input phase after the calculation. Similarly, when the phase is "1", V equals to M11 and R equals to $\overline{M11}$. Note that, when the phases are different, both V and R become high-impedance, so that values in the latches are preserved. Also note that, in every case, both inputs have the same C value. If C is different, the inputs to the LUT will be blocked by the receiver, so that the previous state of the latch will remain unchanged.

C. Interconnection block based on context-ID decoding

We adopt the interconnection block architecture proposed in the paper [6] since it efficiently reduces the area by 50%. The main idea behind this architecture is the redundancy in the configuration data. Since more than 90% of the configuration data are redundant [5], they can be implemented by a simple switch structure smaller than the conventional multi-context switch. Fig.10 (a) shows the implementation of a switch when the configuration data are independent of the context-ID-bit. Fig.10 (b) shows the implementation when the configuration data depend on a single context-ID-bit (C1). This simplified switch is called a switch element (SE) and it is the basic unit in the interconnection block. A switch block (SB) consists with two SEs aligned vertically and horizontally, so that it can act as a pass gate or a crossbar switch.

Fig.11 shows an example of routing two logic blocks, when the configuration data depend on more than one context-ID-bit. In here, LB1 and LB2 are conneted when the context-ID bits are [1,1] and are disconnected in all the other cases. Therefore, we program all the SBs in the desired path except the one near the LB1 as pass gates that are always on. Then we program the SB near the LB1 as a pass gate controlled by an input "S", which equals to $C0 \cdot C1$, so that we can realize the desired connections for all the 4 contexts. Several SBs are combined to create a context-ID-decoder, which generates the control signal "S" using $C0$ and $C1$ as inputs. A detailed description on-routing is given in the paper [6].

Let us explain how the overall architecture works (Fig.12). At the beginning, context-ID-bits are passed to the LBs through the I/O ports. For example, in an 8-context FPVLI, an LB does not transfer the context-ID-bits until all the 3 bits are received. When all the context-ID-bits are available, the context-ID-decoder can create the paths to the adjoining LBs. When a path is created, the LB sends the context-ID-bits through the newly created path to the other LBs that are conneted. After all the paths in the circuit are created, the data can be send through those paths. In the proposed

architecture, the context-ID-bits and the data bits are pipelined to minimize the context switching delay. To run a new context, the appropriate context-ID-bits are passed to the LBs, so that the desired circuit is created automatically. The scheduling of contexts, allocation and mapping should be done in advance in the high-level-synthesis process.

III. EVALUATION

The proposed MC-FPVLSI architecture is designed using 6-metal 1-poly 90nm CMOS design process. The layout of the chip is shown in Fig.13. The designed chip has 100 cells in a 10 × 10 cell array. Since the fabrication process of the actual chip is not completed yet, we have done the HSPICE circuit simulation for two near-by cells. According to the simulation, the minimum data transfer and the context ID transfer delays (for a single bit) are 1.32ns and 0.81ns respectively. Table 1 shows the performance of the chip.

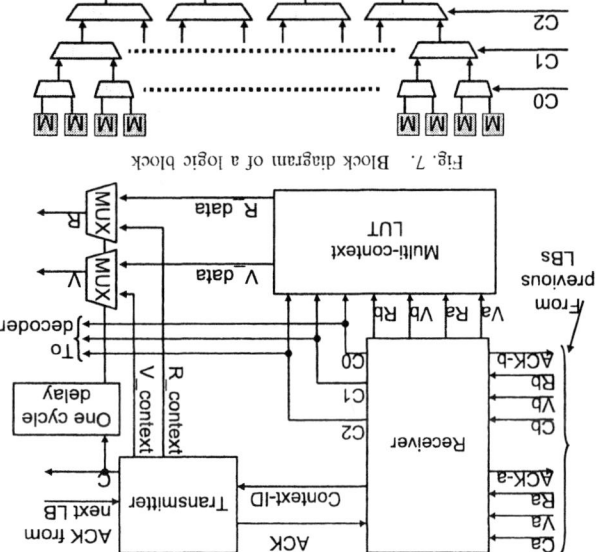

Fig. 7. Block diagram of a logic block

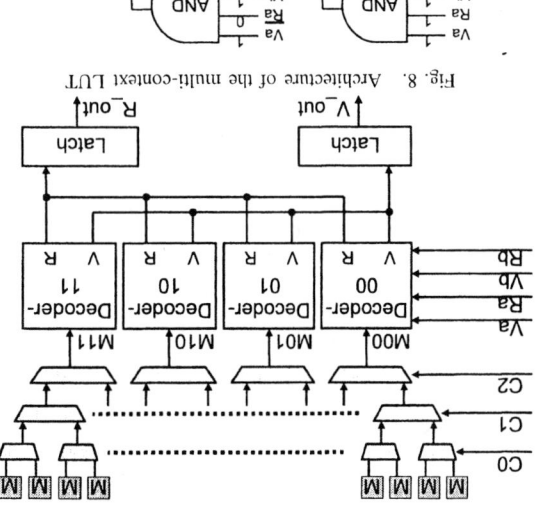

Fig. 8. Architecture of the multi-context LUT

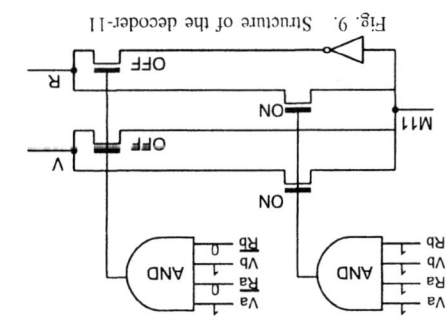

Fig. 9. Structure of the decoder-11

IV. CONCLUSION

We have proposed a novel MC-FPVLSI based on an asynchronous bit-serial architecture. Since local context-ID signals are used, the additional hardware needed to run large number of contexts is very small, unlike the conventional one, where it needs additional context-ID trees when the number of contexts are increased. Therefore, the proposed architecture is better suited for the circuits with large number of contexts.

Not only the area overhead, but also the hardware utilization is better in our architecture, since it can run multiple contexts simultaneously. Note that, when the number of contexts is high, the execution time and hardware utilization for each context varies, so that, even the most powerful high-level-synthesis tools may not be able to achieve a high utilization ratio in conventional MC-FPVLSIs.

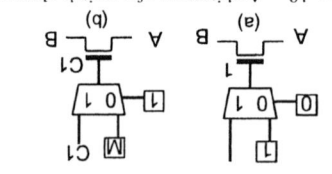

Fig. 12. Overall architecture

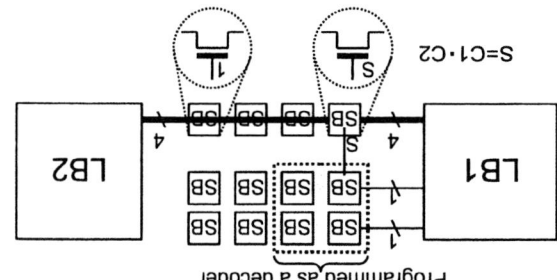

Fig. 11. Routing in proposed MC-FPVLSI

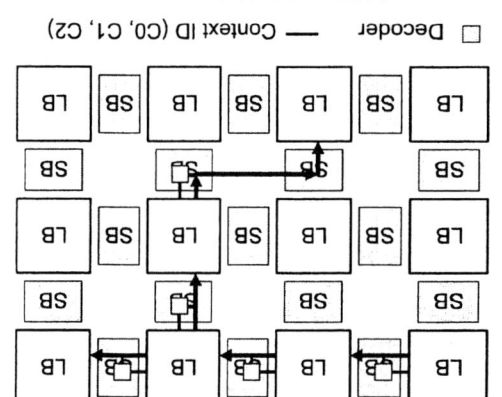

| Context ID | | Connection between LB1 & LB2 |
C1	C0	
0	0	Disconnected
0	1	Disconnected
1	0	Disconnected
1	1	Connected

Programmed as a decoder

$S = C1 \cdot C2$

Fig. 10. Architecture of a switch element

The area of the proposed logic block is larger than that of the conventional one. However, since we are using an interconnection structure based on the context-ID decoding [6], the area of the interconnections is reduced to 50%. Note that, the area of the interconnections accounts for more than 70% of the overall area in a MC-FPVLSI, so that the total area reduction is quite high.

Since we are not using any global clock or global context-ID signal, the peak current is considerably low, compared to the conventional one, where all the cells operate with the rise of a clock or a context-ID signal. As a result, low cost packages can be used in our design.

ACKNOWLEDGMENT

This work is supported by VLSI Design and Education Center (VDEC), the University of Tokyo in collaboration with STARC, Fujitsu Limited, Matsushita Electric Industrial Company Limited, NEC Electronics Corporation, Renesas Technology Corporation, Toshiba Corporation, Cadence Design Systems Inc and Synopsys Inc.

REFERENCES

[1] A. Dehon, "Dynamically Programmable Gate Arrays: A Step Toward Increased Computational Density", Proc. the Fourth Canadian Workshop on Field-Programmable Devices, pp.47-54, 1996.

[2] W. J. Bainbridge and S. B. Furber, "Delay insensitive system-on-chip interconnect using 1-of-4 data encoding", Proc. International Symposium on Advanced Research in Asynchronous Circuits and Systems, pp.118-126, March 2001.

[3] Mark E. Dean, Ted E. Williams and David L. Dill, "Efficient self-timing with level-encoded 2-phase dual-rail (LEDR)" Proc. 1991 University of California/Santa Cruz conference on Advanced research in VLSI, pp.55-70, 1991.

[4] C. Traver, R.B Reese and M.A. Thornton, "Cell designs for self-timed FPGAs" Proc. 14th Annual IEEE International ASIC/SOC Conference, pp.175-179, 2001.

[5] I. Kennedy, "Exploiting redundancy to speedup reconfiguration of an FPGA", Proc. 13th International Conference on Field Programmable Logic and Application, pp. 262-271, Sep. 2003.

[6] Weisheng Chong, M. Hariyama, and M. Kameyama, "Novel switch-block architecture using reconfigurable context memory for multi-context FPGAs", Proc. International Workshop on Applied Reconfigurable Computing (ARC 2005), pp.99-102, 2005.

TABLE I
PERFORMANCE OF THE PROPOSED MC-FPVLSI

Number of contexts	8
Number of cells	100 (10 × 10)
Cell area	53 × 62 (μm^2)
Minimum context ID transfer delay (for single context-ID bit)	0.81 ns
Minimum data transfer delay	1.32 ns

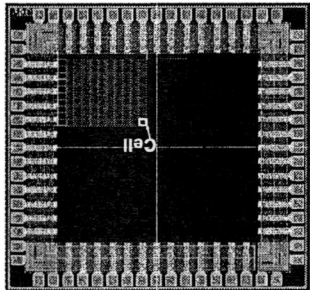

Fig. 13. The layout of the test chip

A Practical Step Forward Toward Software-Defined Radio Transmitters

Essam Atalla*, Imran Bashir†, Poras Balsara*, Kamran Kiasaleh*, Robert Bogdan Staszewski†

*Department of Electrical Engineering, University of Texas at Dallas
†Texas Instruments Inc., Dallas, Texas

Abstract—After many years of research and development in the wireless communication community, software defined radio (SDR) is no longer an unachievable dream. In this paper, we present a significant step forward toward practical SDR transmitters. Motivated by the reconfigurability, programmability and available computational power in a commercial Digital RF Processor (DRP^{TM})-based single-chip GSM/EDGE radio, we succeeded to modulate the RF carrier with P25 compliant C4FM (continuous 4-level FM) data. P25 is a digital public safety standard that operates in the 746-806 MHz frequency band, which is different from the normal operation band of the GSM/EDGE chip. The modulation is based completely on software without need for any hardware modifications. The measurement results show that the transmitted signal spectrum is compliant with the P25 standard specifications. We also show that the work presented in this paper can be extended to provide more elaborate modulation schemes.

I. INTRODUCTION

In the last few years, there has been great development in semiconductor technology that allowed RF and digital baseband integration in a complete system-on-chip (SoC). This was accompanied with the development of wireless communications standards such as WLAN, WiMAX. In addition, cellular communications standards have been modified to provide services beyond voice communications such as high data rate and video communications. For example, EDGE and GPRS are two enhancements that provide high data rate capability for the GSM standard that was originally designed for voice communications. The huge end user demand for integrating different services on a single handset, in addition to the requirement of global roaming, led the manufacturers to strive to develop software-defined radio (SDR).

Despite the great development in semiconductor technology, circuit design techniques are still not capable of providing a full spectrum of SDR operation. That's why different terminology appeared, like multiband, multistandard or multimode, and all are considered under the umbrella of SDR, although they are really not [1]. A transceiver can be referred to as SDR if its communication functions are realized as programs running on a suitable processor. Based on the same hardware, different transmitter/receiver algorithms, which usually describe transmission standards, are implemented in software [2].

In this paper, we will describe an example of a real SDR experience. A commercially available single-chip GSM/EDGE phone based on a Digital RF Processor (DRP^{TM}) technology [3] [4] is software programmed to provide P25 public safety standard transmission, without any hardware modification.

Section II gives an overview of the P25 standard and the GSM/EDGE transmitter that we used. Section III describes the software modulation. The measurement results are given in Section IV. Finally, conclusions are given in Section V.

II. P25 STANDARD AND THE GSM TRANSMITTER

A. Overview of P25 Standard

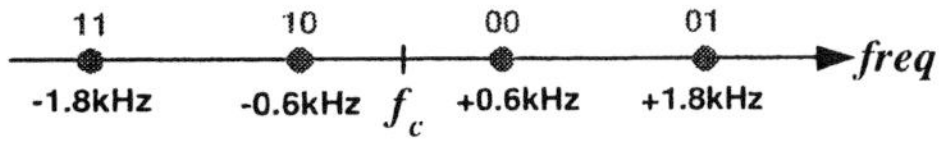

Fig. 1. C4FM modulation in frequency plane showing symbol mappings.

P25, which stands for Project 25, is an open architecture, user driven suite of system standards that define digital radio communications system architectures capable of serving the needs of Public Safety and Government organizations. The P25 suite of standards involves digital Land Mobile Radio (LMR) services for local, state/provincial and national (federal) public safety organizations and agencies [5]. A P25 radio is any radio that conforms to the P25 standard in the way it functions or operates. That is why there is great interest in integrating P25 capability in cellular handsets. P25 compliant technology is being deployed in several phases. Phase 1 uses continuous 4-level FM (C4FM) non-linear modulation for digital transmission in a 12.5 kHz channel. Phase 2 uses CQPSK modulation to transmit digital data over a 6.25 kHz channel. Both Phase 1 and Phase 2 use 9.6 kbps data rate. This is translated into 4.8 kHz symbol rate (2 bits per symbol). In this paper, we focus on Phase 1 only. Figure 1 shows the 4 frequency deviations corresponding to the 4 data symbols.

Phase 1 P25 transmitter employs root raised-cosine pulse shaping filter before modulating the RF carrier. Since the filter shaping filter decides the pulse shape, cannot be realized as infinite in time, we have to choose the length of the filter impulse response (in number of symbols) that will provide proper filtering action in the frequency domain. Proper filtering action means that the frequency response of the filter satisfies the noise profile of the P25 standard. Figure 2 shows the frequency response of the root raised-cosine pulse shaping filter having impulse response length of 2, 3 and 5 symbols. It is quite clear that 2 symbols are sufficient. Yet, using 3 symbols is better because it provides some extra margin.

B. Description of the GSM/EDGE Radio Chip

The GSM/EDGE radio IC that we use in this work is a commercially available chip that has extensive reconfigurability and programmability features. In this paper, our focus is mainly on the transmitter [3]. Figure 3 shows a simplified block diagram for the GSM transmitter. The AM modulation path of the polar transmitter, which is used in EDGE transmission, is not shown in this block diagram, since in this Phase I work we are only interested in the C4FM frequency modulation of the P25 standard. The core of the transmitter is an all-digital PLL (ADPLL), which is formed of digital phase detection and filtering built around a digitally-controlled oscillator (DCO). The digital phase detection circuit provides an integer phase resolution by accumulating clock edges of the frequency reference (FREF) and RF signals and performing a fixed-point subtraction. In order to account for the frequency difference between the FREF and RF signals, the input to the accumulator clocked by the retimed FREF clock (CKR) signal is a frequency control word (FCW). The FCW is the ratio between the desired RF frequency and the reference frequency. The time-to-digital converter (TDC) provides fractional (i.e., sub DCO period) timing resolution.

The phase error at the output of the phase detection and the TDC is filtered in a high order digital filter and fed to the DCO as a frequency tuning word. Frequency modulation is done by adding a modulation frequency word to the FCW. The value of the modulation frequency word is equal to the required frequency deviation from the carrier. For example, a digital word that corresponds to a maximum frequency deviation of ±67 kHz in the GSM modulation.

The modulation data word is read real-time from a random-access memory (RAM) that is clocked at the GSM data rate equal to 270.833 kHz. This means that a new modulation data word is read every 3.692 μsec. Before being added to the FCW at the ADPLL input, the modulation data word is filtered in a Gaussian interpolative filter that provides pulse shaping as defined in the GSM standard [6].

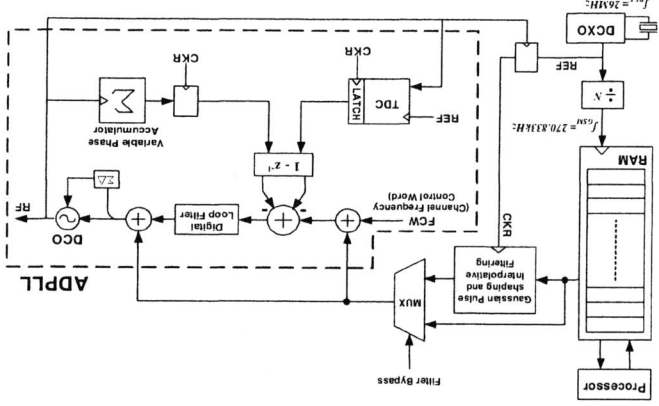

Fig. 3. All-digital PLL based GSM transmitter.

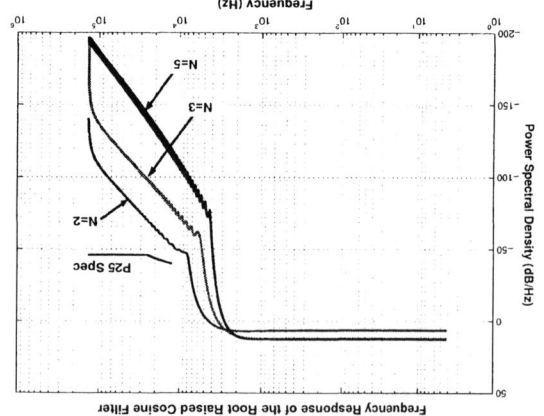

Fig. 2. Effect of the length of pulse shaping filter impulse response.

Figure 4 shows simulation results for the phase noise contributions of the different ADPLL components at its output. It also shows the total phase noise at the ADPLL output. Although the PLL bandwidth is 30 kHz, which is relatively wide corresponding to the P25 maximum frequency deviation of 1.8 kHz, the total ADPLL phase noise is far below the P25 spec. This guarantees that the P25 modulation can be done using this chip. This is indeed very encouraging, since it means that the P25 standard can be integrated into GSM cellular handsets with almost no additional cost.

III. SOFTWARE MODULATION

In the GSM/EDGE transceiver chip, there are embedded processors that perform different configuration and control functions. For example, one task of one of the processors is to set the default values for all the registers in the chip, such as the registers that store the default values of the ADPLL loop filter parameters. One of the embedded processors is shown as part of Figure 3. It is very important with respect to the focus of this paper because it controls the modulation process. It loads the modulation data control word into the RAM. After that, the data word is filtered and applied to the ADPLL to modulate the RF carrier. The best way to

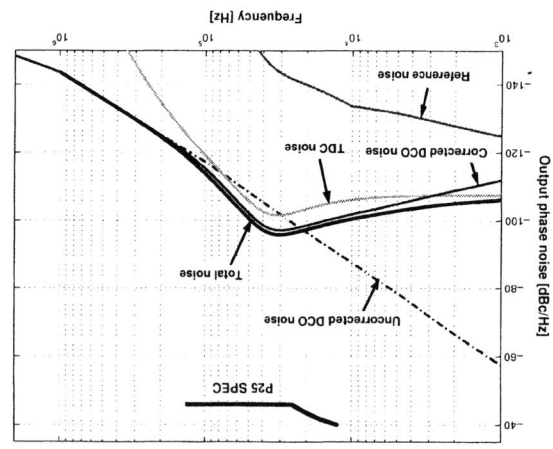

Fig. 4. ADPLL phase noise components at the nominal loop filter settings.

implement a software radio transmitter is to tap the powerful computational power [7] such as DSP and/or ARM processors which are already available on the SoC. While bypassing the GSM gaussian pulse shaping filter (by controlling the MUX shown in Figure 3), a processor can be programmed to handle different modulation schemes. For example, in order to generate P25 C4FM modulation data, the embedded processor should read digitized audio data from a CODEC chip at its input interface, and then perform software root raised-cosine filtering. The filtered data is then written to the RAM and applied directly to the ADPLL (added to FCW). But in this paper we use a slightly different approach. Instead of relying completely on the embedded processor, we used Matlab to generate random data, map the bits to symbols, perform software root raised-cosine filtering, and finally generate the corresponding frequency modulation data word. The results are saved in a file. The embedded processor loads the file into the RAM (Fig. 3) and then writes the frequency modulation data words one-by-one to the ADPLL input. In this way, the output of the ADPLL is P25 compliant C4FM modulated RF carrier. It should be worth mentioning here that the memory pointer, which refers to the memory location from which the modulation data word is read, is configured in a loop-back mode. This means that if we have 100 memory locations storing modulation data words, the pointer will loop back to location 1 after reading from location 100. This guarantees continuous real-time modulation, but the 100 data words must be selected carefully to avoid any periodicity in the output and to guarantee P25 compliant transition from the last symbol (location 100) to the first symbol (location 1). Figure 5 shows a chart that summarizes all the steps in order to perform C4FM modulated carrier.

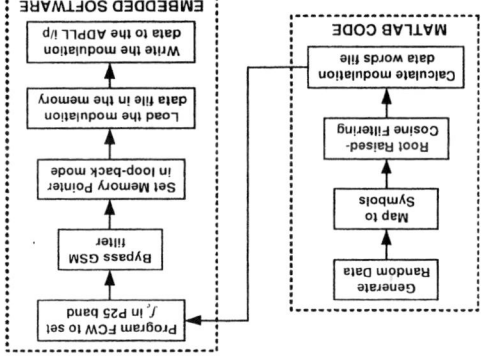

Fig. 5. Steps for performing P25 modulation on GSM chip

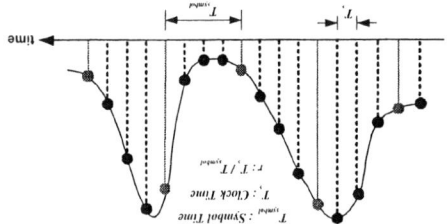

Fig. 6. Illustration of pulse shaping filter action with OSR $r = 4$

A. Limitations of this Approach

The first limitation in the approach we explained above is the memory-read clock frequency. The root raised-cosine pulse shaping filter upsamples the 4.8 kHz symbols, and produces samples at a rate equal to $r \times 4.8$ kHz, where r is the oversampling ratio. Figure 6 illustrates this concept assuming $r = 4$. The modulation data words are read from the memory at the GSM symbol rate equal to 270.833 KHz (because the chip was originally designed for GSM standard). This means that if the ratio between the GSM symbol rate and the P25 symbol rate is x, then each P25 (pulse-shaped) symbol must be represented by x GSM samples. Therefore, the best thing to do is to design the root-raised cosine pulse shaping filter with oversampling factor $r = x$. The problem appears when x is a fractional number. One solution is to use interpolative pulse shaping filter as suggested in [6]. Another solution is to downsample the P25 symbols by a factor D and then use oversampling factor I in the filter, such that the net oversampling ratio $r = I/D$ is equal to the non-integer ratio x.

In this paper, for the sake of concept verification we approximated the ratio $x = 270.833/4.8 = 56.423$ is non-integer, although the filter upsampling factor r to 57. The described solutions are based on the fact that the memory read clock is fixed and equal to the GSM symbol rate as shown in Figure 3. A better solution is to have a programmable memory-read clock that can be chosen such that the ratio x is always an integer. This solution is not available in the specific chip we were using, since it was designed for GSM/EDGE applications. The second limitation in the approach we used is the memory size. Assuming that the memory size is n locations, the maximum number of P25 symbols that can be stored in the memory is equal to n/x where x is the number of GSM samples per P25 symbol rounded to the nearest integer. In our experiment, we had 1000 memory locations available, and so we were able to store 17 P25 pulse-shaped symbols. And these symbols are repeated in the loop-back mode as described in the last section.

IV. MEASUREMENT RESULTS

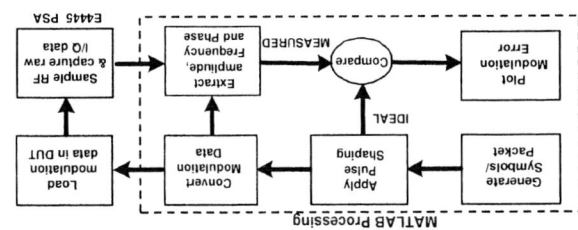

Fig. 7. Validation flow of P25 Phase 1 standard

Devising a common platform for system verification and testing of radios supporting diverse order of modulation schemes is a challenge. In most cases, the instruments are designed to support a specific set of most commonly used wireless standards in the commercial sector. A more efficient approach to this problem, which was used in this project, is to use a generic vector spectrum analyzer as a sampler to output raw demodulated data. The modulation characteristics of the RF signal can then be determined by post-processing the raw data. This flow is graphically represented in Figure 7.

P25 C4FM modulation performance is analyzed using the stated flow. First, symbols are generated in MATLAB script from a random stream of bits. The data is then passed through a raised-cosine pulse-shaping filter with an upsampling factor of 56, roll-off factor of 0.25, and filter group delay of 4. This filter simulates the combined effect of the two root-raised cosine filters used in the transmitter and receiver. The DRP modulator has a fixed sampling clock of 270.833 kSps and when an integer upsampling of 56 is applied, the resulting data rate at transmitter output is 4.836 kSps.

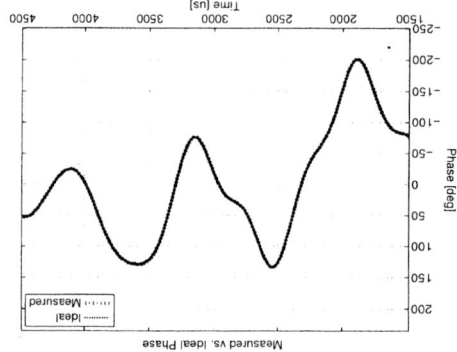

Fig. 8. Ideal and measured phase trajectory vs. time

Figure 8 shows the simulated and measured phase trajectory over the transmitted packet. The simulated phase is obtained from the MATLAB script and measured phase is determined from raw I/Q data from the vector spectrum analyzer.

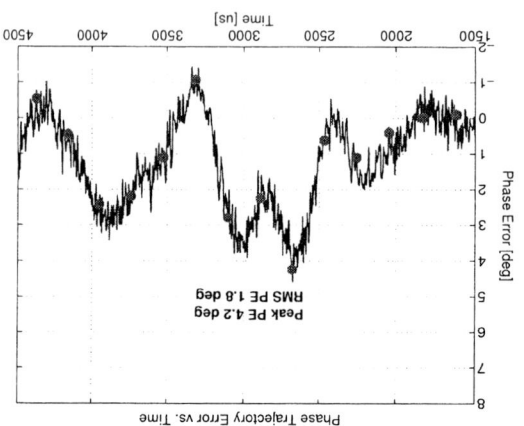

Fig. 9. Phase trajectory error vs. time

Figure 9 shows the phase trajectory error which is the numerical difference of ideal and measured phase over the transmitted packet. The red circles on the plot mark the position of the 14 symbols in the packet. The RMS phase error is calculated by taking the root-mean-square of the phase error measured at the symbols shown in the plot which is around 1.8°. Similarly, the peak phase error is the maximum phase error over the symbols which is 4.2°. Figure 10 represents the spectrum measured while C4FM modulation was applied to the transmitter. The spread of the carrier due to modulation is not symmetric. This is due to the fact that the data packet of 14 symbols is repeated over time and hence the data is not entirely random during the measurement. The phase noise at 12.5 kHz offset from the carrier, where the adjacent channel is located, is -62 dBc.

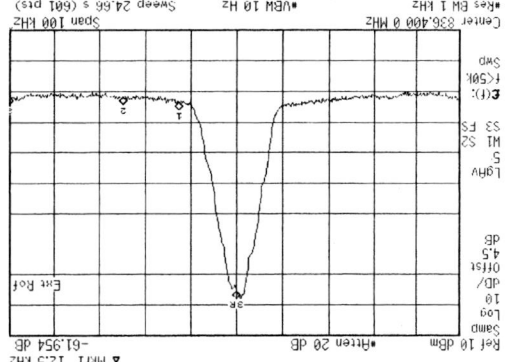

Fig. 10. P25 C4FM modulated spectrum

V. CONCLUSIONS

We have presented a practical realization of a software-defined radio (SDR) transmitter based on a GSM single-chip phone using the Digital RF Processor (DRP^{TM}) technology. The high reconfigurability and software programmability of the DRP transmitter allowed us to meet specifications of the Phase-1 P25 public radio standard.

ACKNOWLEDGMENT

The authors would like to thank National Institute of Justice (NIJ) for funding this research project.

REFERENCES

[1] R. Bagheri, A. Mirzaei, S. Chehrazi, et al., "An 800MHz to 5GHz software-defined radio receiver in 90nm CMOS," in Proc. of IEEE Solid-State Circuits Conf., 2006, pp. 480-481.

[2] F. Jondral, "Software-Defined Radio - Basics and Evolution to Cognitive Radio," EURASIP Journal on Wireless Communications and Networking, no. 3, pp. 275-283, 2005.

[3] R. B. Staszewski et al., "All-Digital PLL and Transmitter for Mobile Phones," IEEE Journal of Solid State Circuits, vol. 40, no. 12, pp. 2469-2482, Dec. 2005.

[4] K. Muhammad et al., "The First Fully Integrated Quad-Band GSM/GPRS Receiver in a 90-nm Digital CMOS Process," IEEE Journal of Solid State Circuits, vol. 41, no. 8, pp. 1772-1783, Aug. 2006.

[5] Daniels Electronics LTD. (2004) P25 training guide. [Online]. Available: http://www.p25.com/resources/P25TrainingGuide.pdf

[6] R. B. Staszewski and R. Staszewski, "Interpolative Pulse-Shape Filtering for a GSM/Bluetooth Transmitter," in IEEE Dallas Circuits and Systems Workshop (DCAS), 2005, pp. 191-194.

[7] R. Staszewski et al., "Software Assisted Digital RF Processor for Single-Chip GSM Radio in 90 nm CMOS," in IEEE Custom Integrated Circuits Conference (CICC), 2006, pp. 611-614.

Dual-Threshold Voltage Technique for Asynchronous Pre-Charge Full Buffer Linear-Pipelines

Behnam Ghavami   **Hossein Pedram**

ghavamib@aut.ac.ir, pedram@ce.aut.ac.ir

Computer Engineering Department, Amirkabir University of Technology (Tehran Polytechnic)
424 Hafez Ave, Tehran 15785, Iran

Abstract

Scaling the technology and reducing the feature size in integrated circuits have caused leakage power consumption to become one of the main challenges to the digital design. Dual-threshold CMOS circuit, which has both high and low threshold transistors in a single chip, can be used to deal with the leakage problem in high performance applications. This paper presents dual-threshold voltage technique for reducing leakage power dissipation of Pre-Charge Full Buffer asynchronous linear-pipelines while still maintaining high performance. We employed Folded Dependency Graph to produce a formal performance analysis. In order to reduce leakage power an algorithm for assigning a high threshold voltage is proposed. Results obtained indicate that our proposed technique can achieve on average 30% savings for leakage power, while there is no performance penalty.

Keywords

Asynchronous Circuit, Leakage Power, Dual-Vt, PCFB.

1. Introduction

Reduction in the size and the growth in number of the transistors in contemporary circuits signify the problem of global synchronization. One solution is to eliminate the global clock signal and take advantages of asynchronous design methods. As asynchronous circuits gain popularity due to their potential advantages, such as dynamic power saving and high performance, the complexity of design and synthesis methods is highlighted. Among the numerous asynchronous design styles being developed, template-based have decreased the complexity of design effort [1].

As of today, high performance digital design is formidably challenged by high power consumption. In the sub-micron regime, leakage currents make up a significant portion of the total power consumption in high-performance digital circuits [2]. In asynchronous circuits as one class of VLSI circuits, leakage power increases with the scaling of CMOS manufacturing technology into deep sub-micron era. Hence, designers require techniques that reduce leakage power while maintaining high performance of these circuits.

Main component of the leakage power is due to the subthreshold leakage current and it is becoming an increasingly dominant component of overall power consumption in deep sub-micron technologies [3]. There is a lot of researches around employing subthreshold leakage power reduction techniques in synchronous circuits [2][4].

In this paper, we have introduced an efficient methodology for employing dual-threshold voltage techniques in Pre-Charge Full Buffer (PCFB) asynchronous linear-pipelines. We proposed the assignment of low threshold voltage (low-VT) and high threshold voltage (high-VT) to the transistors in templates of pipeline, in such a way that the delay remains the same as in the case of all low-VT designs, but reduces subthreshold leakage current significantly. The remainder of this paper is organized as follows. We discuss about Quasi Delay Insensitive (QDI) asynchronous linear-pipeline in Section 2. Section 3 is a review over background of dual-Vt technique and some discussions. Then Section 4 elaborates the proposed methodology for dual-Vt asynchronous linear pipeline. Section 5 is about our experimental results in detail by the use of some related test-benches. Finally, some conclusions are drawn in Section 6.

2. QDI Asynchronous linear-Pipeline

Asynchronous circuits represent a class of circuits not controlled by a global clock but rely on exchanging local request and acknowledge signaling for the purpose of synchronization. An asynchronous circuit is called delay-insensitive (DI) if it preserves its functionality, independent of the delays of gates and wires[5]. It is shown that the range of the circuits that can be implemented completely DI is very limited. Therefore some timing assumptions exist in different design styles that must hold to ensure the correctness of the circuit. Different techniques distinguish themselves in the choice of the compromises to the delay-insensitivity. Quasi delay insensitive (QDI) circuits are like DI circuits with a weak timing constraint [5]. QDI asynchronous circuits are composed of concurrent processes connected through handshake channels. These processes can be decomposed into fine grain processes that all fit in in a fine grained template.

At present, most QDI pipelines are designed using pre-charged half buffer (PCHB) and pre-charged full buffer (PCFB). This paper discusses the idea of leakage power reduction in the context of QDI asynchronous linear PCFB pipelines.

Linear-pipeline is defined as pipeline structure that contains no fork or join. A template at $stage_n$ becomes active when it senses the presence of an incoming data from function block of $stage_{n-1}$. It then performs the computation and sends the result via output channels to $stage_{n+1}$. Communications through channels are controlled by handshake protocols. One of the major protocols used in asynchronous circuits is four-phase protocol [5]. Buffer Cycle Time, C_B, is the time needed by the buffer at stage n to process a complete data pattern. To isolates the pipeline from the environment effects, we assume that the starting of the pipeline (TX) produces a new token as soon as it has the acknowledgment and ending point of pipeline (RX) produces the acknowledgment as soon as it has a new token. Figure 2 shows a PCFB linear pipeline [8].

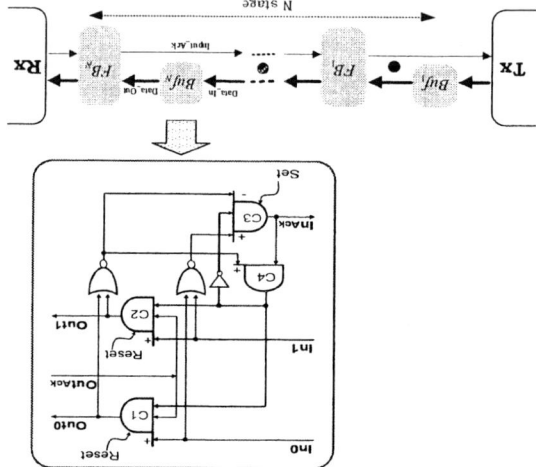

Figure 1. PCFB linear-pipeline

3. Dual-Threshold Voltage CMOS: Background and Discussion

Researchers have proposed different circuit techniques to reduce subthreshold leakage power of synchronous circuits [2].

Multiple-threshold CMOS circuit, which has both high and low threshold transistors in a single chip, can be used to deal with the subthreshold leakage problem in low power and high performance applications. The high threshold transistors can suppress the subthreshold leakage current, while the low threshold transistors are used to achieve the high performance. Recently several multiple-threshold CMOS circuit design techniques have been provided.

In dual-threshold technique, a higher threshold voltage can be assigned to some transistors in non-critical paths so as to reduce leakage current, while the performance is maintained due to the low threshold transistors in the critical path(s). Figure 1 illustrates the basic idea of a dual-Vth scheme in synchronous circuits.

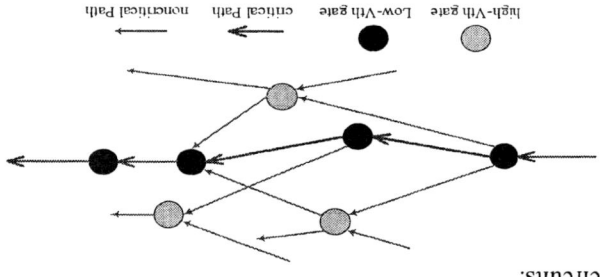

Figure 2. Dual Vth scheme in synchronous circuits

Since dual-threshold design technique can reduce both active leakage power and standby leakage power without delay and area overheads, it is very attractive for low voltage and high performance circuit design. However, due to the complexity of a circuit, not all the transistors in non-critical paths can be assigned a high threshold voltage; otherwise, the critical path may change, thereby increasing the critical delay.

Recently, researchers have proposed many design techniques, for selecting and assigning an optimal high threshold voltage to gates of synchronous circuits which reduce leakage power under performance constraints [2]. But in case of asynchronous circuits, applying these techniques has serious problems. This is due to the fact that estimation and analysis of the performance of asynchronous circuit remains somewhat of a stumbling block because of dependencies between highly concurrent events [6][7].

While synchronous performance estimation is based on a static critical path analysis affected only by the delay of components and interconnecting wires, it has been shown that the performance estimation of asynchronous circuit is more complex [6]. In asynchronous circuits, the operation of a system proceeds at a rate determined by speed of its individual components, and the sequencing of the operation of components. The techniques required to analyze asynchronous systems resemble those used to determine the clock period of a synchronous system, which is, summing the delays along the longest path through the combinational logic connecting adjacent latches. In the clocked case, the critical path has a *clear beginning* and a *clear end* because all paths are broken by latches. But importantly, no clear separation is available in asynchronous circuits. Analysis procedures must deal directly with *cyclic critical paths*; thus, existing critical-path analysis tools cannot be easily applied to this problem.

In brief, traditional dual-Vth techniques can not be employed directly to asynchronous circuits. So, in next sections, we first introduce asynchronous linear pipelines, and then an abstract performance model of this circuits on which the dual-Vt problem can be applied is proposed. Non-linear pipelines are under investigation.

4. Dual-Vt PCFB linear-pipelines design methodology

To employ Dual-Vt technique, at first a suitable performance model for PCFB linear pipelines is presented, and then an efficient algorithm for assigning high and low Vt to components of PCFB pipeline is proposed.

4.1 Formal Performance-Analysis

For analyzing the performance of PCFB pipeline, a formal and detailed equation is needed. We need exact equation to assign the suitable Vt for their circuits without affecting the performance. In order to determine the cycle time of a pipeline, it is necessary to analyze the dependencies of the required sequence of transitions. These dependencies can be drawn in a marked directed graph where the nodes of the graph correspond to specific rising or falling transitions of circuit components, and the edges represent the dependencies of each transition on the outputs of other circuit components. The delay of each transition is represented by a value attached to the corresponding node in the graph. These graphs will be called "Dependency Graphs" [8][9].

If all the stages have the same function blocks, the graph can be folded. Each edge in the Folded Graph is annotated with an integer weight giving the offset in stage indices to which that dependency refers. Cycles in the Folded Graph whose edge weights sum to zero correspond to the cycles in the original Dependency Graph and thus the zero-weight cycle with the largest sum of node delay values gives the cycle time. More details can be found in [8]. Figure 3 shows the folded dependency-graphs of the PCFB Pipeline.

Figure 3. The Folded Dependency Graph for the PCFB pipeline

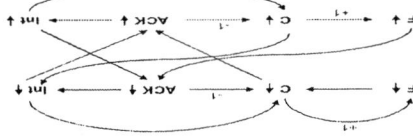

The possible loops of PCFB pipeline are, (F↑, C↑), (F↓, Ack↑, Int↑, Ack↓, C↓) with index of (+1) and (Ack↓, C↑, Int↑), (Ack↑, C↑, Ack↓, F↓), (Ack↑, C↑, Ack↓, Int↑), (Ack↓, C↑, Int↑), with index of (-1). Finally, the loop (F↓, Ack↑, Int↑, Ack↓, C↓) has an index of (0). The resultant cycle time equation is shown in eq.1.

$$C_{PCFB} = Max [(T_{F\downarrow} + T_{Ack\uparrow} + T_{Int\uparrow} + T_{Ack\downarrow} + T_{C\downarrow}),$$
$$(Max[T_{F\downarrow} + T_{C\uparrow}), (T_{F\downarrow} + T_{Ack\downarrow} + T_{Int\uparrow} + T_{C\downarrow}$$
$$+ T_{Ack\downarrow} + T_{C\uparrow} + T_{F\downarrow}), (T_{Ack\downarrow} + T_{Int\uparrow} + T_{C\uparrow}$$
$$+ T_{Int\uparrow} + T_{C\uparrow}))]. \tag{1}$$

The propagation delay through $node_x$ denoted as $T(x)\uparrow$ or $T(x)\downarrow$, defines how quickly the output responds to a change in the input. Using RC-delay model, relation of propagation delay by value of Vt is resolved [4].

4.2 Dual-VT Assignment

We propose the assignment of low threshold voltage (low-VT) and high threshold voltage (high-VT) to the transistors in PCFB template in such a way that the delay remains same as in case of all low-VT design, but reduces sub-threshold leakage current significantly. Circuits are comprised linear and non-linear pipelines. In this paper only linear pipelines are considered. The first step of this algorithm is to initialize a circuit with a single low threshold. The low-threshold is determined by the performance requirement. After initialization, all delay parameters associated with each node are computed. Using eq.1, the cycle time of the pipeline is determined. At this step, all elements that don not participate in the critical loop are resolved and assigned to high-Vt. After updating the network for high-Vt, the parameters of circuit are updated. The pseudo-code for the initialization procedure is shown below.

5. Experimental Results

The method to reduce leakage power using dual-threshold-voltage transistors has been implemented in C under the Berkeley SIS environment. All the simulation results were obtained using HSPICE with the BSIM3V3 model for a 0.18μ MOSIS process. The effective channel length of the transistor is taken as 0.18μm and the gate oxide thickness is taken as 40Å. For simplicity, all transistors are assumed to have the same channel length of 0.18μm, while the channel widths for nMOSFETs and pMOSFETs are assumed to be 0.54μm and 1.62μm, respectively. The sub-threshold swing coefficient (γ) is taken as 1.44 and the body effect coefficients (η) and DIBL coefficients (δ) are 0.03 and 0.21 for nMOSFETs and 0.02 and 0.11 for pMOSFETs, respectively. For the active mode and standby mode of circuit, temperatures are assumed as $110^\circ C$ and $25^\circ C$, respectively. The supply voltage is assumed 1.0V and zero biased low threshold voltage and high threshold voltage used in our experiments is 0.2V and 0.5V, respectively.

```
Assign-Vt () {
1. for each pipeline of circuit, DO
{
2. Calculate the propagation delay Tphl(x), Tplh(x) of each node x.
3. Determine the possible loops
4. Calculate the C_PCFB of pipeline.
5. Until there is unmarked nod repeat
{
6. Update network with assign high-Vt to one node and mark it.
7. Update the new propagation delay Tplh(x), Tplh(x) of updated node.
8. Update the delay of loops.
9. Calculate the new C_PCFB of pipeline.
10. If new C_PCFB <= C_PCFB): go to step 5.
Else: remove high-Vt from last node and go to step 5
}
}
}
```

Persia[10] is a QDI synthesis toolset that is employed to synthesis our benchmarks. In order to feasibility of similar functional units in a pipeline, folded dependency graph model, technology-mapping was used to map the circuits to a library which contains NAND gates. We have tested our approach with benchmark and experimental results are presented in Table 1. Leakage power in active and standby mode of the circuits with single-VT and dual-VT realizations is shown. It is observed that, on the average, in dual-VT PCFB circuits 36% and 24% leakage power can be reduced in active mode and standby mode, respectively.

Table 1: Active and standby leakage power savings for dual-Vth PCFB circuits

Circuit	Gate #	Leakage power (uW) in active mode			Leakage power (uW) in standby mode		
		Single-VT	Dual-VT	%	Single-VT	Dual-VT	%
DiffEq	548	311.4	214.8	31	17.4	13.74	21
GFAdder	358	195.3	140.6	28	9.6	7.77	19
HammingEnc	503	291.7	177.9	39	15.9	11.76	26
Syndrome	625	330.9	215.0	35	20.1	14.47	28
ChienForney	807	461.8	290.9	37	26.5	15.9	40
RiBM	1103	627.1	407.6	35	34.1	26.25	23
ReedSolomon	2505	1561.8	1030	34	73.4	52.84	28

6. Conclusion

Reduction in leakage power has become an important concern in low power and high performance applications. In this paper, we introduced dual-threshold design technique for asynchronous PCFB linear pipeline. Formal performance analysis using the folded-graph method has been done. In order to reduce leakage power under performance constraints starting with a single low Vth circuit, an algorithm for assigning a high threshold voltage is proposed. Results show that both active and standby leakage power can be reduced by more than 30% for some of the circuits.

7. References

[1] A. M. Lines "Pipelined Asynchronous circuits" MSc Thesis, California Institute of Technology, June 1995, revised 1998

[2] Amit Agarwal, Saibal Mukhopadhyay, Arijit Raychowdhury, Kaushik Roy, Chris H. Kim. "Leakage power analysis and reduction for nanoscale circuits", ieee 2006.

[3] S. Borkar, "Design Challenges of Technology Scaling," IEEE Micro, vol. 19, no. 4, July-Aug. 1999, pp. 23-29.

[4] S. Mutoh et al., "1-V Power Supply High-Speed Digital Circuit Technology with Multithreshold Voltage CMOS," IEEE J. Solid-State Circuits, vol. 30, no. 8, Aug. 1995, pp. 847-854.

[5] Jens Sparso, Steve Furber, "Principles of Asynchronous Circuit Design – A System Perspective", Kluwer Academic Publishers, 2002.

[6] Sangyun kim and peter a. beerel. pipeline optimization for asynchronous circuits: complexity analysis and an efficient optimal algorithm. IEEE Trans. on computer-aided design of integrated circuits and systems, vol. 25, no. 3, march 2006

[7] C. V. Ramamoorthy and G. S. Ho, "Performance evaluation of asynchronous concurrent systems using Petri nets," IEEE Trans. Softw. Eng, vol. 6, no. 5, pp. 440-449, Sep. 1980.

[8] Eslam Yahya, M.Renaudin, QDI Latches Characteristics and Asynchronous Linear-Pipeline Performance Analysis, Research Report, TIMA-RR--06/-01--FR (2006).

[9] Ted Williams: Performance of Iterative Computation in Self-Timed Rings, Journal of VLSI Signal Processing, 7, 17-31 (1994).

[10] Persia Site: http://www.async.ir/persia/persia.php

LAYOUT PARASITIC INTERCONNECTIONS EFFECTS ON HIGH FREQUENCY CIRCUITS

Cristian M. Albina, Member IEEE, and Günther Hackl

GME mbH, 82008 Unterhaching, Germany

ABSTRACT

The rapid growth of microelectronics constantly presents new challenges to the IC designer. The physical and dynamic characteristics of wires on a die begin to dictate the topology of an integrated circuit. Second- and third-order effects are becoming important in designs built on processes smaller than 400 nm. In this paper we try to present the influence of the parasitic layout elements by showing the difference between RC and RLC parasitic extraction and simulation and their effects on the performance of a limiting amplifier used in the optic fiber transceivers. The evaluation was done using a standard 150 nm CMOS 6 metals technology.

Index Terms — Interconnects, LIA, RLC, transmission line

1. INTRODUCTION

Until some years ago, gate delays represented the most significant portion of a signal-propagation time. Things changed with the new sub-micron technologies [1] of which most the important parameters are illustrated in Table 1.

Table 1. Selected Data from the 2005 SIA Roadmap

Year	2003	2006	2009	2012
Transistors	79M	200M	520M	1.4G
CLK (GHz)	2.1	3.5	6.0	10.0
Layers	7	7-8	8-9	9
Chip size (mm^2)	430	520	620	750
Power dissipation(W)	130	160	170	175

Wire delays became more important than gate delays. Gates are now smaller than most of the interconnecting wires (Figure 1). In addition, the calculation of the delay that wiring causes must take into account second- and third-order parameters. Long wires exhibit now similar effects to the transmission lines with faster transistor rise and fall times. The inductance component of the wires becomes comparable to their resistance component leading to effects like signal ringing and inductive crosstalk between the signal paths additional to the traditional parasitic resistance and capacitive elements. Circuit delays become critical as timing requirements become tighter. All this on-chip parasitic effects require proper extraction and modeling methods in order to provide the designer with adequate tools for a successful and reliable design. During the past years several methods where presented for the on-chip crosstalk modeling [2] and [3]. In our paper we'll show the influence of the physical layout components on the overall performance of a transinductance and limiting amplifier for optical fiber communication systems, insisting on the gain and phase margin values and the output eye diagram.

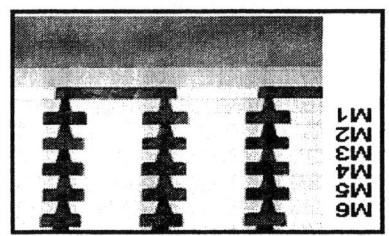

Figure 1. CMOS Technology Layers Stack

2. RLC PARASITIC EXTRACTION AND MODELING

An electrical transmission line can be modeled as a two-port network, as follows (Figure 2):

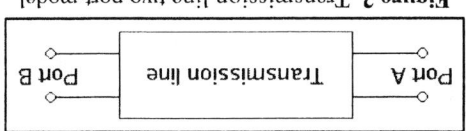

Figure 2. Transmission line two port model

For the simplest simulation case we can assume that the network is linear and that the two ports are interchangeable. If we consider that the transmission line is uniform along its length then the behavior can be modeled by one single parameter called the characteristic impedance Z_0. Typical values are 50 or 75 Ω for the coaxial cable or 100 Ω for a twisted pair of wires. The transmission line is then modeled like an infinite series of elementary two-port components each representing an infinitesimally short segment of the transmission line. When loss is significant, the effects of the series resistance (R) and the dielectric conductance (G) should be included like in Figure 3.

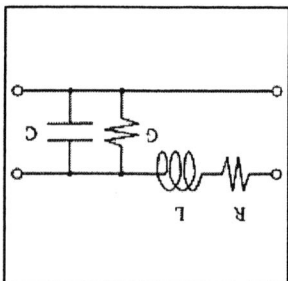

Figure 3. Equivalent circuit model for the transmission line

The interconnects inside the integrated circuits have an important contribution through their R, L and C components on the electrical circuit parameters like delay, crosstalk or matching, especially in deep sub-micron technologies and that's why it becomes more important to take these effects into account during the IC simulations. Unfortunately, after running the parasitic RC or RLC extraction of the circuit based on the physical implementation, the size of the netlist is in most cases too big to be simulated because during the extraction all the parasitics for all geometrical structures which were found are taken into account. The process of eliminating the extracted circuit negligible elements is called netlist reduction [4] and it can sometimes reduce the simulation time by a factor up to 10 without loosing the simulation accuracy. Most of the algorithms developed use as a main parameter the cut-off frequency and try to eliminate the internal nodes which have no effect within the given frequency range for the circuit. If the cut-off frequency of the node is much higher than the cut-off frequency of the circuit then we can eliminate this node (1).

$$f_{signal} << f_N = 1/T_N \qquad (1)$$

For our evaluation we'll use a simplified model of the transmission line without the component G.

3. INTERCONNECTION PARASITIC EFFECTS ON THE CIRCUIT PERFORMANCE

To illustrate the effects of the parasitic elements of the layout on the circuit behavior we'll use a limiting amplifier LIA used in the integrated transceivers for optic fiber communications. The layout was implemented using the Mixed-Mode TSMC CMOS 0.18 µm T-018-MM with 4 metal layers stack. The requirements for the TIA/LIA chains were:

- Supply voltage 3-3.45 V
- Gain 40dB minimum
- Differential output voltage 800mV
- 1.25Gbps with 2.5Gbps option
- Sensitivity 4 mV
- Input impedance 100 Ω
- Output resistance 50 Ω single ended
- Jitter 10-20 ps
- PSSR 35 dB

Preamplifiers for gigabit-per-second transmission systems are realized as TIA's (transimpedance amplifier) which transform the small output current of the photodiode into an output voltage of, e.g., some millivolts. The main advantages compared to high-impedance amplifiers are the higher dynamic range and the elimination of the critical equalizing circuitry [5], [6]. Figure 4 shows a block diagram of a standard optical receiver architecture used for long and short haul optical communication systems. To implement the LIA we will use several cascaded trans-impedance operational amplifiers TIA which allows us to achieve a higher gain.

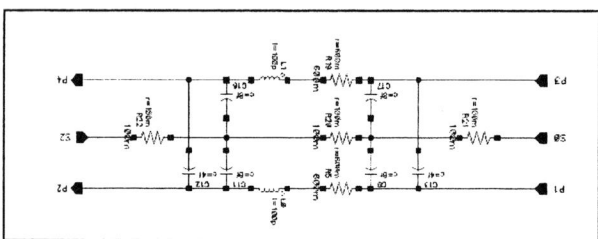

Figure 4. Optical Receiver Block Diagram

For the lossy transmission line between the LIA's we will use the following schematic as shown in Figure 5 taking into account the RLC elements of the interconnects.

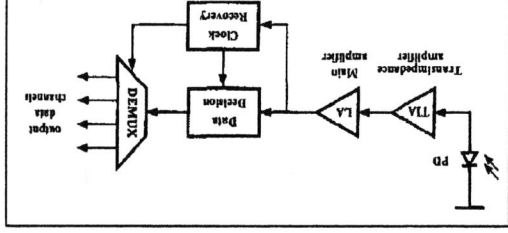

Figure 5. Transmission line RLC simulation model

In our model we included the parasitic capacitances of the wires towards VSS and the cross-coupled parasitic capacitances resulted after the layout extraction as well as the parasitic resistances of the signal wires and of the supply lines between each buffer stage of the LIA.

For deep-submicron, high-performance circuits, ignoring inductance effects may incur a large amount of error, since an RC model as compared to an RLC model may create errors of up to 30% in the total propagation delay of a repeater system. Inductance causes overshoots and undershoots in the signal waveforms, which can adversely affect signal integrity. For global wires inductance effects are more severe due to the lower resistance of the lines, which makes the reactive component of the wire impedance comparable to the resistive component, and also due to the presence of the mutual inductive coupling between wires due to the long return current paths. For our example we calculated the parasitic inductance based on the method presented by Banerjee [7]. For submicron technologies the values for the line inductance are between 1.8 and 2.2 nH/mm. Each LIA was implemented using the cascaded fully differential amplifier circuit (Figure 6).

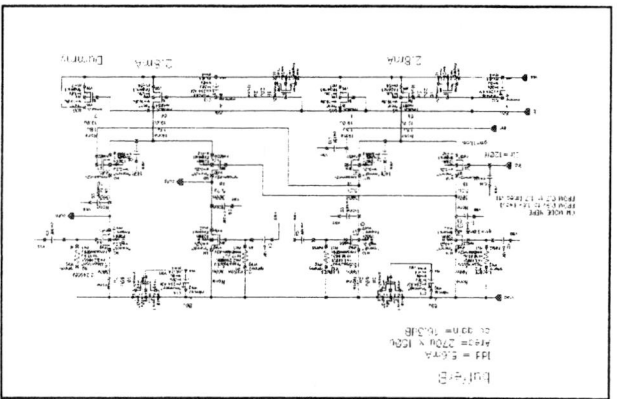

Figure 6. Limiting amplifier buffer schematic

For the circuit parasitic elements extraction we used the layout from Figure 7.

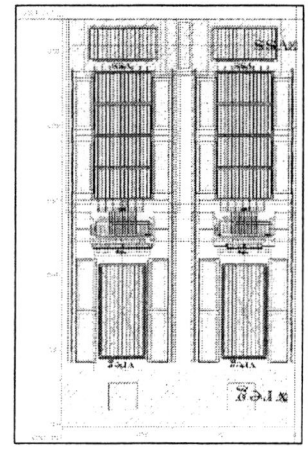

Figure 7. Limiting amplifier buffer layout

3.1. RC parasitic extraction simulation results

The circuit simulations without taking into account the parasitic effects of the layout show us the following values after each stage: 1st 18,3 dB/-33 deg, 2nd 36.31 dB/-68 deg, 3rd 54, 31 dB/-71deg (Figure 8).

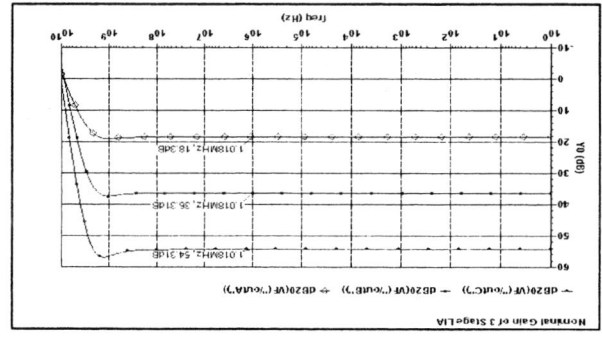

Figure 8. LIA circuit gain simulation

If we add the RC parasitic values extracted from the layout as well as the temperature influence over the circuit performance we can see a variation of the overall gain for the 3 stages with almost 5 dB between 59.64 dB and 54.8 dB at the interest frequency of 1.25 GHz and a reduction of the noise margin with almost 30 degrees from -71 to -106 degrees (Figures 9, 10).

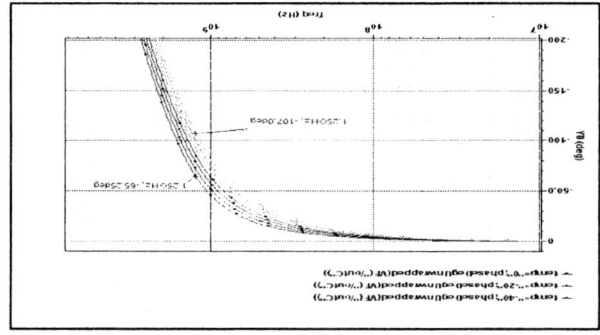

Figure 9. LIA RC extracted circuit gain simulation

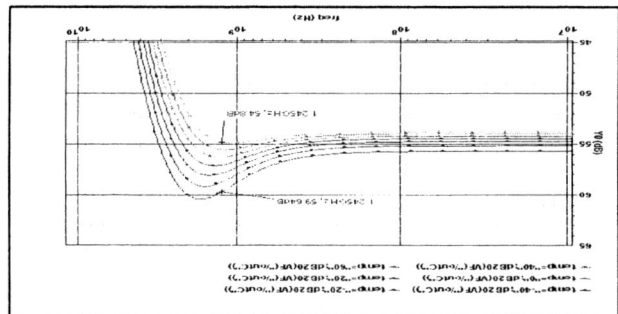

Figure 10. LIA RC extracted circuit noise margin simulation

3.2. RLC parasitic extraction simulation results

Redoing the simulations taking this time into account the complete RLC transmission line model and the full circuit layout we obtained an overall variation of the gain of 7 dB due to the L component of the circuit as well as the coupling factor between two adjacent lines (Figures 11, 12, 13):

4. CONCLUSIONS

For critical circuits it is always necessary to do a full simulation including the parasitic effects of the physical layout. As the working frequencies increase is better to take into account also the on-chip inductance effects of the wires on the signal integrity and the power and ground noise. It's becoming important to have accurate 3D geometry models and analytic formulae for self and mutual inductance estimation. The overshoot and under shoot effects can cause catastrophic failures both in terms of device life-time degradation and errors during the operation of the logic circuits. Designs in sub-micron technologies are more susceptible to inductance effects. That's why for a critical circuit is always necessary to do a full simulation based on the full RLC-parameter extraction even if it is time and resources consuming. To put it in a nutshell, no matter what technology is considered crosstalk remains a huge problem.

5. REFERENCES

[1] International Technology Roadmap for Semiconductors 2004 Edition, http://public.itrs.net.

[2] B. A. Floyd et al. "Intra-Chip Wireless Interconnect for Clock Distribution Implemented with Integrated Antennas, Receivers and Transmitters", IEEE-JSSC Vol. 37, No. 5, pp. 543-552, 2002.

[3] X. Bai, S. Dey, "High-Level Crosstalk Defect Simulation Methodology for System-on-Chip Interconnects", IEEE Transaction on Computer-Aided Design of Integrated Circuits and Systems, Vol.23, No. 9, pp. 345-349, 2004.

[4] C.M. Albina, G. Hackl, "Layout parasitic limitations in high speed circuits", IEEE Circuits and Systems, International Semiconductor Conference, CAS 2006, Vol.2, pp. 375-378, 2006.

[5] S. Galal et al. "10 Gb/s limiting amplifier and laser/modulator driver in 0.18 um CMOS technology" IEEE-JSSC Vol.38 No. 12, pp. 2138-2146, 2003.

[6] E. A. Crain, M. H. Perrot "A 3.125 Gb/s limit amplifier with 42 dB gain in 0.18 um CMOS" IEEE-ISSCC Dig. Tech. Papers, pp. 232-234, 2005.

[7] K. Banerjee, A.Mehrotra, "Analysis of On-Chip Inductance Effects using a Novel Performance Optimization Methodology for Distributed RLC", Proceedings of DAC, pp.798-803, June 2001.

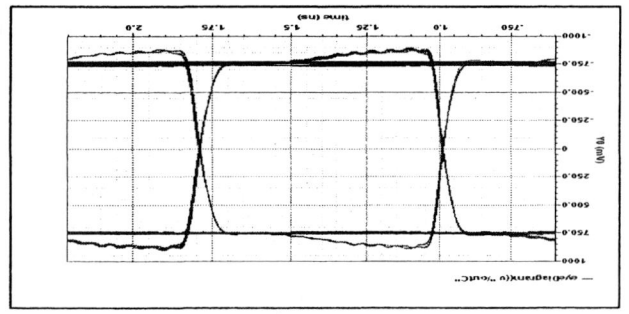

Figure 15. RLC extracted circuit eye diagram

The signal ringing effects can be visualized also in the signal eye diagram. The measured eye opening varies between 1.5 and 1.8 volts with stronger variations if we include the inductive crosstalk between the lines (Figures 14, 15).

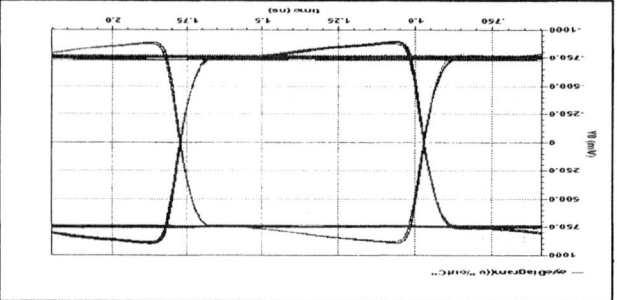

Figure 14. RC extracted circuit eye diagram

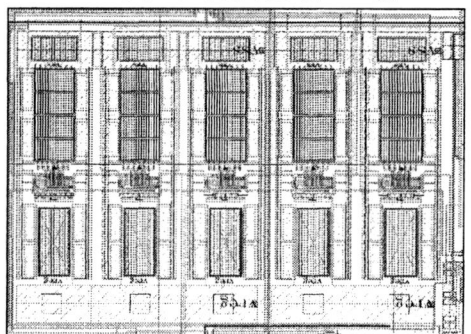

Figure 13. TIA-LIA circuit layout

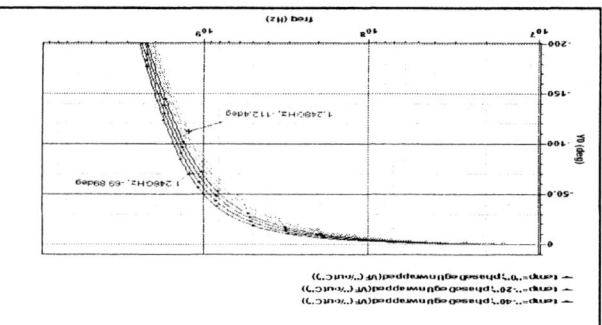

Figure 12. LIA RLC extracted circuit noise margin simulation

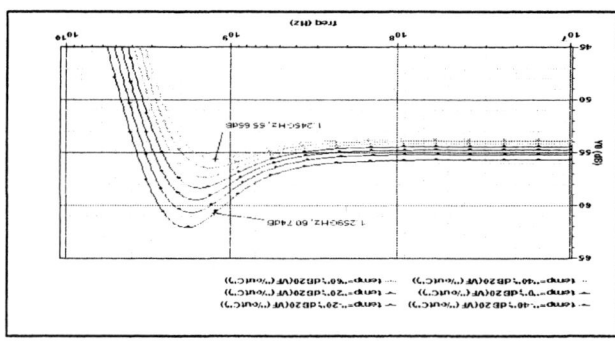

Figure 11. LIA RLC extracted circuit gain simulation results

"Flying-Adder" PLL Based Synchronization Mechanism for Data Packet Transport

Liming Xiu, Steve Clynes, Srikanth Gurrapu, Towfique Haider, Feng Ying, Wahed Mohammed

Texas Instruments Inc, Dallas, Texas, USA

Abstract—A novel frequency synthesis architecture, Flying-Adder architecture, has been developed in recent years. Compared with conventional PLL based frequency synthesis techniques, this new method has many unique features. Among them, the two most distinguishing ones are its instantaneous response speed and very fine frequency resolution. These features can be especially beneficial to application of data smoothing in packet-oriented transport systems. In this paper, this Flying-Adder PLL based synchronization approach is demonstrated through two real examples.

Index Terms--PLL, Flying-Adder, data packet, USB, MPEG2, data synchronization, data smoothing, isochronous.

I. INTRODUCTION

In many electronic systems, packet oriented transmission is used to pass information from one device to others. These systems include USB bus, 1394 (FireWire) interface, HDMI interface, MPEG2 transport system and etc. Among these devices, the transmitter sends the data packets using its clock and, at the receiving end, the packets are received synchronously on an independent local clock. In some cases, the clock signal is not transmitted along with the data. In others, the clock signal is available but the data is not presented in every clock cycle. In both cases, the transmitter and/or receiver clock sources can drift over a long period of time and cause loss or duplication of data. This is undesirable for real-time audio/video transmission such as USB audio speakers or video conferencing. Therefore, certain synchronization mechanism has to be established at the receiver side for smooth operation in downstream processing, as shown in figure 1.

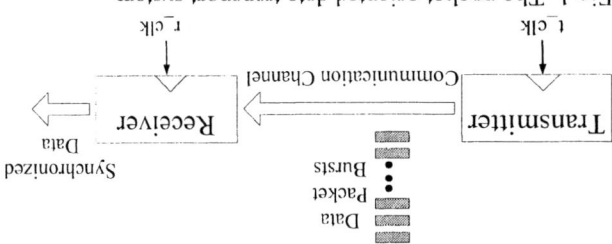

Fig. 1. The packet oriented data transport system.

Inside the receiver, First-In-First-Out (FIFO) memory is usually employed to accommodate the different data rates at

the two ends as shown in figure 2. Traditionally, PLL is used to generate the output clock (r_clk). Based on the fullness/emptiness of the FIFO memory, the PLL output clock frequency is adjusted so that there is continuous data flow out of the receiver at constant rate. Clearly, the PLL's response speed is critical in this feedback mechanism. For slow PLL, which requires significant amount of time to respond to the FIFO's status change, the size of the FIFO memory has to be sufficiently large to avoid data loss. Another issue is the frequency resolution. Ideally, the output clock frequency should be adjustable slightly around its center value for small input data rate variation. However, conventional PLL has difficulty in achieving fine frequency resolution. Consequently, the frequency change is more or less abrupt.

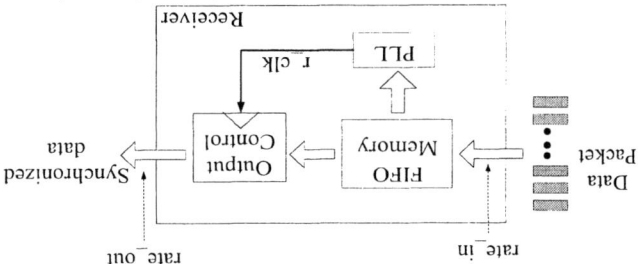

Fig. 2. The synchronization mechanism inside the receiver.

Flying-Adder frequency synthesizer is ideal for such applications owing to its instantaneous response speed and very fine resolution [1][2]. It is a circuit level enabler which enables system level novel solutions for this data packet smoothing application. Section II of this paper is the brief description of the Flying-Adder architecture. Section III is the demonstration of using Flying-Adder PLL to function as a stream controler for USB audio speaker. Section IV shows how the Flying-Adder PLL is used in MPEG2 data transport system for reducing memory resource requirement. Section V concludes the paper.

II. BRIEF DESCRIPTION OF FLYING-ADDER SYNTHESIZER

A. The principle idea

The "Flying-Adder" frequency and phase synthesis architecture is a technique of producing desired frequency from a group of equally-spaced VCO outputs. The idea is conceptually shown in figure 3. The VCO is locked to a crystal reference through a PLL. The synthesizer generates the desired frequency by triggering the toggle-configured

Liming XIU, Texas Instruments Inc., limingxiu@ti.com

978-1-4244-1679-0/07/$25.00 ©2007 IEEE

D-type Flip-Flip at predetermined time through the selection of different VCO outputs. The output frequency is controlled by a frequency control word FREQ. The detail circuit implementation can be found in [1].

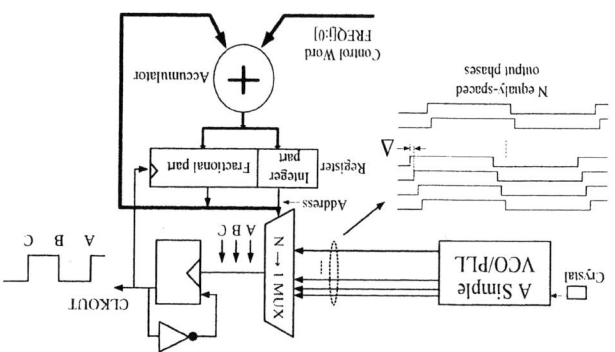

Fig.3. The principal idea of Flying-Adder architecture.

The typical Flying-Adder PLL implementation is shown in figure 4. Its operating equation is:

$$f_o / f_r = (N * K) / (FREQ * P * M) \quad (1)$$

Where f_r is the input reference, P is the pre scalar, N is the PLL loop divider and M is the post divider. K is the number of VCO outputs. In fixed-VCO Flying-Adder mode, the VCO oscillation frequency is fixed with P and N preset to fixed values. Usually, the input reference f_r is a known and fixed value. Thus, the output frequency f_o is only dependent on FREQ when post divider M is also fixed. In this operating mode the frequency transfer function becomes (2) where $C = (N*K*f_r)/(P*M)$ is a constant.

$$f_o = C / FREQ \quad (2)$$

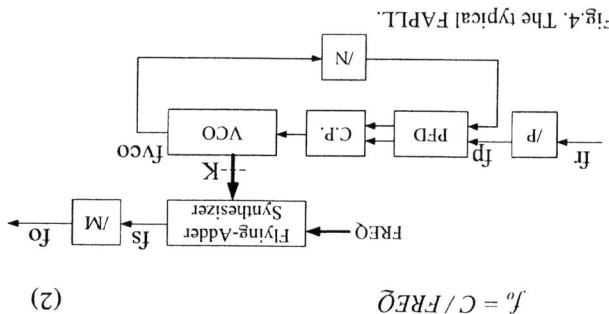

Fig.4. The typical FAPLL.

B. Instantaneous response speed

Whenever there is a FREQ update, the synthesizer's output frequency will be correspondingly changed in next clock cycle. This is owed to the fact that the VCO is always running at a fixed frequency and the synthesizer directly constructs the output clock's waveform (period) for generating the desired frequencies. Detail circuitry can be found in [1].

C. Fine frequency resolution

The resolution can be expressed in (3).

$$\delta f = -2^{-p} * \Delta * f^2 \quad (3)$$

Where p is the number of fractional bits in FREQ. f is the synthesizer's output frequency. δf is the frequency step at this frequency. Detail analysis can be found in [2].

III. USB STREAM CONTROLLER

Fig.5. is the simplified block diagram of a USB audio speaker system. The USB peripheral interface is used to transfer the digital audio data from PC to a speaker by isochronous streaming. The CODEC is used to convert the digital data into analog signal which then drives the speaker. The most commonly used sampling rates are: 22.05 KHz for lower-quality PCM and MPEG audio, 44.1 KHz for audio CD and 48 KHz for DVD. However, sending the isochronous USB data directly to CODEC/speaker could be problematic. In the case of 48 KHz, the isochronous USB mode only guarantees that there is a burst packet consisting of 48 audio samples between two Start Of Frames (SOF, USB terminology). The placement of this packet could be anywhere within 1 ms time frame. In other words, The 48 KHz data sent out by the PC (asynchronous) is not necessarily aligned with the 48 KHz clock (synchronous operation) used in CODEC. This could result in some erratic audio with Clicks and Pops which is very evident if not compensated. The problem is even more complex for other sample rates. For example, at 44.1 KHz, the USB host controller sends 9 44-sample bursts followed by 1 45-sample burst in a repetitive pattern. It is the responsibility of the USB device to provide this as a continuous stream at the 44.1 KHz rate to the speakers.

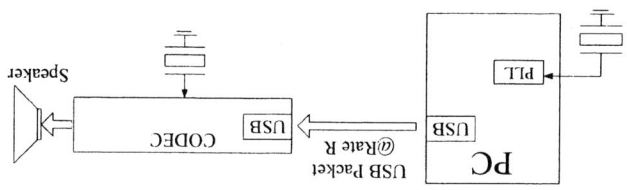

Fig.5. PC and USB speaker.

Fig.6. is the solution to this problem by inserting a stream controller between the PC and the speaker [3]. The function of the stream controller is to take the isochronous USB data and smooth them, or synchronize them, before sending them to the CODEC/speaker. The controller stores the data in its memory after receiving them from the USB interface. The center frequency generated by the Adaptive Clock Generator (ACG) is 48 KHz (in the case of 48 KHz sampling rate) which is used to drive the audio CODEC. Moreover, as shown, the ACG also receives the USB data. Based on the time stamp (SOF) embedded in the USB packet, the ACG can slightly adjust its output frequency to accommodate the variation in the USB data stream so that audio noise can be avoided. While making these on-the-fly adjustments, it's also important to keep the jitter to the minimum. Because of the finer accuracy in ACG, the closed loop parameters (step size, response time) can be modified to achieve high fidelity sound.

The ACG block structure is depicted in figure 7. The key component is a Flying-Adder frequency synthesizer. From the time-stamps of SOF embedded in the USB data, the current data rate can be derived. Then, the desired frequency value can be calculated and converted to the frequency control word (FREQ) by software. This FREQ will be applied to the synthesizer and produces the desired frequency for CODEC clock. Due to the instantaneous response speed of Flying-Adder synthesizer, this process can be carried out continuously in real time. Furthermore, since the Flying-Adder synthesizer's frequency resolution is extremely fine, any small variation in the USB data rate can be sensed and followed by the ACG.

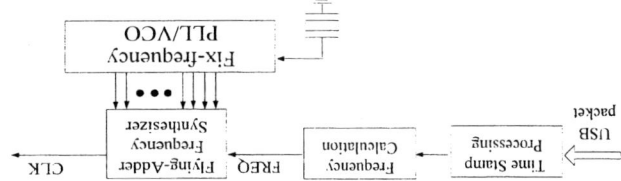

Fig.7. The Adaptive Clock Generator (ACG).

In this stream controller implementation, the Flying-Adder synthesizer is constructed from a VCO of 32 outputs. There are 24 bits in FREQ [23:0] where the top 6 bits are for integer and the reminding 18 bits are for fractional part. The Fix-frequency PLL/VCO is referenced from a 6 MHz crystal. The output CLK is capable of varying in the range from 750 KHz to 25 MHz. The frequency resolution is approximately 4 Hz, thanks to the 18 fractional bits.

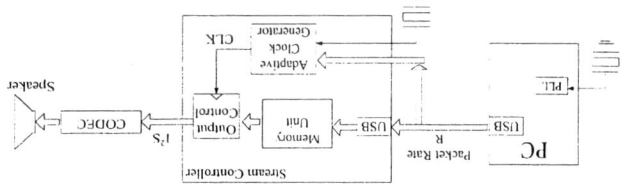

A 1KHz (48kHz sample rate) Tone streaming from the PC to USB Audio Speakers

Fig.8. The tones generated with and without the ACG.

In operation, the USB SOF signal is captured by the clock CLK (or other clock with known frequency) and stored in a register. This stored value is valid until next SOF occurs (about every 1 ms). The on-chip processor can read this FREQ value, do the calculation for new FREQ value, and subsequently update the FREQ control register for applying the new frequency. By locking to the SOF in this fashion, the controller can guaranty the smooth data flow between the PC and the speaker. Figure 8 shows the 1 KHz tones generated by the data sent to the speaker by the PC. As shown, when the USB data is directly sent, there are occurrences of sample dropping/repeating. Whereas, when ACG is used, a clean tone is created.

IV. MPEG2 DATA PACKET SYNCHRONIZATION

MPEG2 transport stream (MPEG2-TS) is a data format specified in part 1, system, of ISO/IEC 13818-1. Its purpose is to allow the multiplexing of digital video and audio data, and the synchronizing of the output. The basic data unit in MPEG2-TS is the packet, which is usually 188 bytes long but could be 204 bytes as well. The data in transport system can be transmitted in either packet synchronous mode or packet asynchronous mode. Figure 9 shows the time diagram of the synchronous mode. In this mode, the clock rate is the same as the data rate. In other words, there is valid data in every clock cycle. Signal DVALID is used to indicate whether the data is valid or not during the current clock cycle. Thus, for synchronous mode, the DVALID is active high all the time. The PSYNC signal indicates the start of the packet.

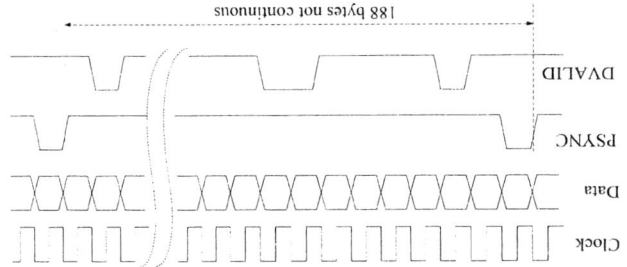

Fig.9. MPEG2 packet synchronous transmit mode.

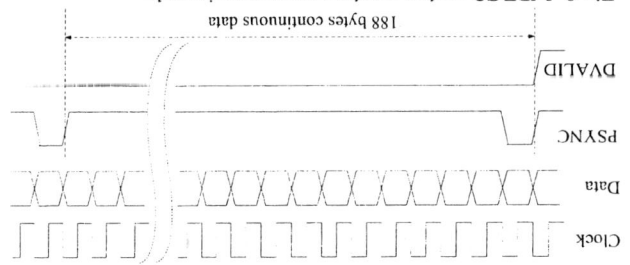

Fig.10. MPEG2 packet asynchronous transmit mode.

Figure 10 is the time diagram of the asynchronous mode. In this mode, the clock rate is higher than the data rate. At each clock cycle, there may or may not be a valid data. The data is valid only when the DVALID is high. Some MPEG2 receivers are only capable of receiving synchronous data. In these cases, the asynchronous MPEG2 data has to be

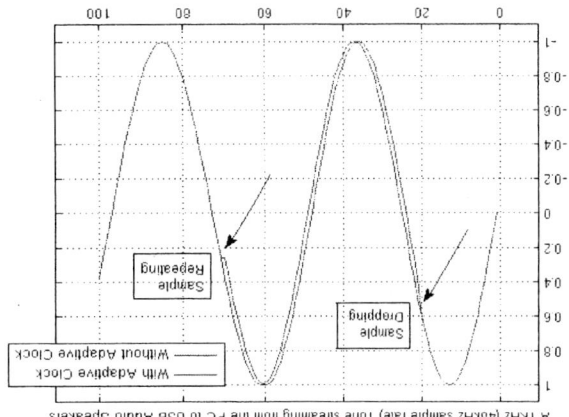

Fig.6. The USB stream controller.

synchronized before being sent to these devices. This process is often referred as packet synchronization or packet smoothing.

Conventionally, the packet synchronizer is realized by the structure depicted in figure 11. Since the input data rate (controlled by write clock) is not the same as the output data rate (controlled by read clock), a FIFO memory device is required in the data path. The clock frequency of the read clock is derived from a PLL which is controlled by the fullness/emphiness of the FIFO and/or the input data rate. The resulting read clock frequency has to match the average data rate of the input data.

When conventional PLL is used, the PLL's response always bears a certain time since the loop bandwidth is constrained by its input reference frequency and loop stability requirement. Usually, the PLL is designed with low bandwidth (slow response speed) to confine the noise from the input side and maintain the high loop stability. Consequently, the read clock cannot follow the input data rate's shift immediately. In order to avoid the possibility of data loss, the size of the FIFO memory has to be large enough so that it can hold all the necessary data when rate shift occurs. If a slow PLL is used, the FIFO size could be as large as several hundreds of packets. In contrast, if Flying-Adder PLL is used, the memory requirement can be reduced to as small size as two packets.

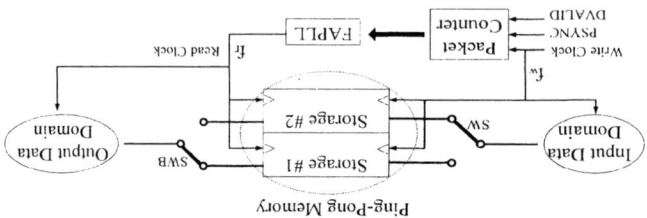

Fig.11. The packet synchronizer using conventional PLL.

Figure 12 shows the idea of using Flying-Adder PLL to achieve the packet smoothing function. In this scheme, there are two storage units employed, each having the size of exactly one packet. They are called Ping-Pong memory since at any given time there is always one unit being read from and one unit being written to, and vice verse.

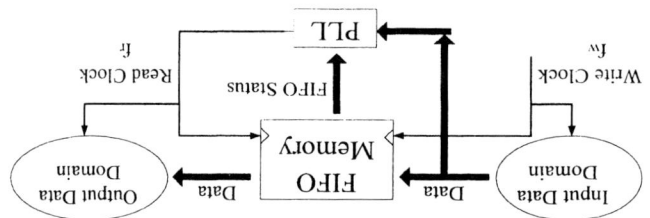

Fig.12. The packet synchronizer using Flying-Adder PLL.

In the input data side which is driven by the higher speed write clock, there is a packet counter that constantly counts the number of clock cycles used for transmitting one packet. The structure of this packet counter is depicted in figure 13. The valid-data-counter is used to count the number of clock cycles which are companioned with valid data, between two consecutive PSYNC pulses. The clock-pulse-counter is used to count the total number of clock pulses between two consecutive PSYNC pulses. At the beginning of any new packet, the PSYNC becomes active high. This signal will reset the two counters. And at the same time, the contents of the two counters will be latched into two registers, register A and B, respectively. The content inside the RegisterB will be compared with a value of Packet Size (PS). If equal (e.g. all are 188 or 204), the Packet Valid (PV) signal will be asserted to indicate the fact that a valid packet has been received. Otherwise, it stays low. The content in RegisterA (N) is the number of total clock pulse used for current packet received/stored in memory. Subsequently, the read clock frequency can be calculated as $f_r = (PS/N)*f_w$. Since Flying-Adder PLL can response instantly, the new frequency value can be applied to FREQ in next read cycle.

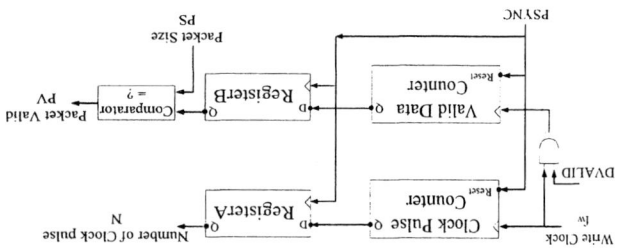

Fig.13. The packet counter.

Figure 14 shows the control states for switch SW and SWB (figure 12). WR1 and WR2 represent the operations of writing to storage #1 and #2, respectively. The incoming packet size can be 240+ or 400+ (valid + redundant data) depending on the standards used. The write state machine filters that out to 188/204 data per packet. RD1 and RD2 are states for reading from them. The symbols of RW1, RW2, RR1, and etc. are the rules for state-changes indicated by the corresponding arcs. Following tables list the content of these rules where PV1 is the PV signal for writing to storage #1. PV1 = 1 when one packet is successfully written to storage #1, PV2 is the same signal but for storage #2.

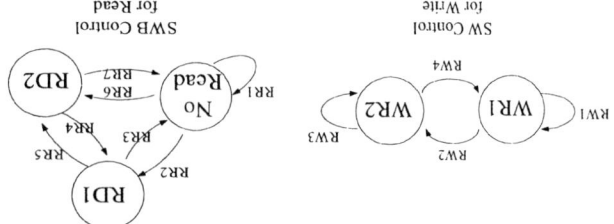

Fig.14. The control states for switch SW and SWB.

Table 1. The control rules for write switch SW.

Rule	Logic
RW1	NOT PV1
RW2	PV1 AND PSYNCH
RW3	NOT PV2
RW4	PV2 AND PSYNCH

Figure 15 shows one simulation of the MPEG2 packet synchronizer. In this plot, the signal *wclk* is the write clock. The *dvalid* indicates the validity of the data. The *wclk_count* reports the number of *wclk* cycles used for the packet. When the write operation is finished, an interrupt (*cpu_intrpt*) to on-chip processor could be generated if necessary. Based on the *wclk_count*, the CPU will calculate a new frequency control word for Flying-Adder PLL to adjust the read clock, *rclk*. The resulting *rclk*, along with the data, will be used in the downstream processing. The scheme described in this section is a precise operation; only two storage units are needed exactly. The memory requirement has been significantly reduced. It has been utilized in a digital TV demodulator chip [4][5].

V. CONCLUSION

In any application that involves the data transfer between devices, there exists the data synchronization, or data flow smoothing problem. At the transmitting side, the data could be sent out in synchronous, asynchronous or isochronous mode. After the data is received at the receiving side, it has to be synchronized relative to certain clock so that downstream processing can be carried out efficiently. Flying-Adder PLL is ideal for implementing this clock-related feedback mechanism since its response speed is instantaneous and its frequency resolution is extremely fine. Compared to conventional PLL based approaches, this new method certainly has the advantages of lower cost and better performance.

REFERENCE

[1] L. Xiu and Z. You, "A 'Flying-Adder' Architecture of Frequency and Phase Synthesis with Scalability", IEEE Transaction on VLSI, pp. 637-649, Oct, 2002.
[2] L. Xiu and Z. You, "A New Frequency Synthesis Method based on 'Flying-Adder' Architecture", IEEE Trans. on Circuit & System II, pp. 130-134, March, 2003.
[3] Data manual, "TAS1020 Data Manual: USB Streaming Controller", SLAS263A, Texas Instruments Inc., 2001.
[4] Ying et al., "MPEG-2 Transport Stream Packet Synchronizer", US Patent, pending, 2006.
[5] Data manual, "TVP9900 VSB/QAM Receiver Data Manual (Rev.A)", Texas Instruments Inc., 2007.

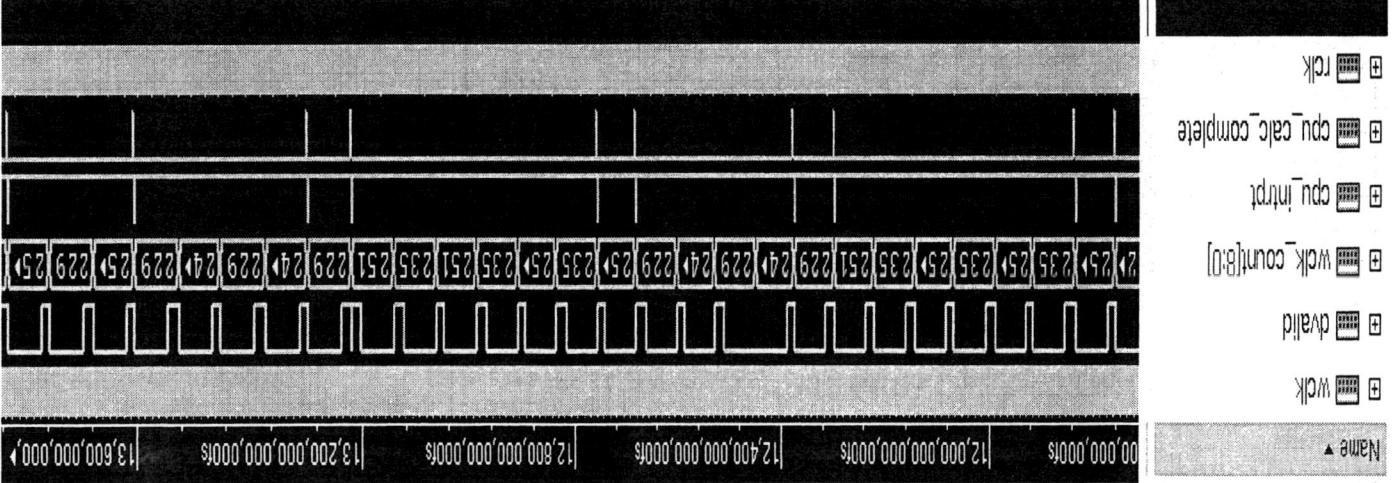

Fig.15. The simulation of MPEG2 packet synchronizer.

Table 2. The control rules for read switch SWB

Rule	Logic
RR1	(NOT PV1) AND (NOT PV2)
RR2	PV1
RR3	(RD1 Done) and NOT PV2
RR4	(RD2 Done) and PV1
RR5	(RD1 Done) and PV2
RR6	PV2
RR7	(RD2 Done) and NOT PV1

SYSTEM-ON-CHIP POWER CONSUMPTION REFINEMENT AND ANALYSIS

David Y. Feinstein, Mitchell A. Thornton and Fatih Kocan
Department of Computer Science and Engineering
Southern Methodist University
Dallas, TX, USA
{dfeinste,mitch,kocan}@engr.smu.edu

Abstract

Accurate power consumption estimation of a System-on-Chip (SoC) using modeling techniques is difficult due to the diverse mixture of processes with radically different current consumption. It is very important that these estimations will be fine tuned to the specific SoC with accurate current measurement during the design and prototyping phase. We introduce an accurate method to measure power consumption using a single measurement point and a dynamic logging algorithm. We present a demonstration tool for continuous logging of the instantaneous power consumption with an identification of the running process within the SoC. Our approach can also be used to steer the dynamic power management (DPM) of a SoC.

1. INTRODUCTION

The emergence of battery-operated System-on-Chip (SoC) in recent years increased the efforts to reduce the power consumption. Early works identified the need of minimizing the power consumption at the initial hardware/software co-design process [4,10]. More current works showed the crucial need for efficient power consumption simulation and estimation tools [2,7,8]. The increased complexity of modern SoC applications limits the capabilities of such simulation tools to predict the exact measured power consumption. In particular, the memory, the processing, and the analog components of the typical SoC have radically different power consumption profiles.

The ability of a SoC system to perform within its power budget is often achieved using dynamic power management (DPM) methods. DPM must employ a reliable real-time means for measuring the actual power consumption [1,5].

This paper discloses a new method to dynamically obtain accurate power consumption measurements of the individual sub-circuits and their associated processes using a *single* *measurement point* within the SoC architecture. The data obtained can be used to fine tune the SoC power simulation tools. When incorporated within the design methodology of SoC, our approach can pinpoint the hardware/software sub-components that require power consumption refining in order to meet the system's power budget. Our approach can be further integrated in the SoC design to augment its DPM method.

2. RELATED WORK

The general "power wall" problem that limits the Moore's law exponential growth of semiconductor density also affects all types of contemporary SoC systems [1,5]. Recent SoC power consumption reduction efforts employ various innovative approaches. Lackey et al. surveyed the Voltage Islands methods that reduce the active and static SoC power consumption [6]. They emphasized the need for accurate current measurement that does not require costly simulation-based switching vectors. Hillman described the Virtual-Silicon's VIP mobilized power management approach which is also based on power islands [5]. State-based power analysis for SoC was offered by Bergamaschi and Jiang [2]. They noted that the generic power consumption used for the power estimation can be very inaccurate if applied blindly without proper "tuning" to the target application. Lee et al. developed the **Power ViP** framework to provide cycle-accurate power estimation for SoC at the transaction level [7]. They took the component-based approach in order to achieve a fast and easy power model. Their work demonstrated the need for accurate actual power consumption verification based on experimental measurement.

Chandrakasan et al. [3] offered the following analysis of the main power consumption contributors in CMOS devices:

$$P_{total} = C_L V_{swing} V_{dd} f_{clk} + I_{sc} V_{dd} + I_{leak} V_{dd} \qquad (1)$$

The first term is the dynamic power consumption. C_L is the circuit capacitance; V_{swing} is the voltage swing which is equal or less to the supply voltage V_{dd}. f_{clk} is the circuit clock frequency, or the switching rate of the circuits. Their paper covered traditional power consumption reduction techniques based on voltage scaling, frequency reduction and leakage current minimization.

Benini et al. surveyed several dynamic power management techniques based on system level considerations [1]. They also indicated the need for an accurate real-time current measurement to dynamically steer the DPM system.

The average current consumption can be easily measured by an ammeter connected between the power supply and the circuit. This method cannot capture the instantaneous power consumption of a SoC where different tasks are performed in a complex order, often for a fraction of a second. A common technique to determine the current consumption of a specific task using a similar arrangement runs the task in an endless loop [12].

Watanbe et al. investigated the use of pipeline task scheduling for power reduction while satisfying both the throughput and the latency constraints [11]. Actual power measurements to support such task scheduling are hard to achieve. The running tasks are often preempted by other tasks so that the measurements must gain insight into the real-time power consumption profile of the application.

This paper is motivated by such needs for accurate power consumption measurements that can improve the SoC simulation tools and improve the power consumption refinement within the design process.

3. OUR APPROACH

We propose a SoC design that allows an accurate real-time measurement and analysis of the power requirements of the various processes *using a single measurement point*. We integrate a low cost current measurement sensor into the SoC architecture and develop a dynamic protocol that determines the *individual power consumption of each process*. This allows the circuit designer to iteratively refine the power consumption of the design by identifying the processes that demand the most power and trace their behavior in the real-time target application. A SoC utilizing our approach provides accurate power consumption measurements to tune up power simulation tools like those developed by Bergamaschi and Jiang [2].

The sensor can often be readily implemented with very minimal cost using the existing resources of the SoC. Although the current measurement circuitry may be removed from the SoC after the design phase, it can be left in the design to help steer the DPM system if implemented in the SoC [1].

It is important to insure that the integrated current measurement process poses a minimal and *fixed* execution time overhead on the SoC. Similarly, the sensor itself and the overhead in performing the current measurement should have a small and *fixed* current consumption overhead. Meeting these two constraints make it possible to "calibrate out" the effect of the measurement process from the overall measurements. We should note that typical existing SoC A/D channel resources include a built-in data averaging capability that minimizes the overhead.

Since the SoC runs numerous tasks that may preempt each other, it is crucial to identify the tasks that are associated with each power measurement. We offer the protocol shown in Fig. 1. The Current Log Routine periodically reads the average current consumption and logs this reading together with the time ticker. This process is repeated indefinitely based on a time interrogation or other polling techniques.

Each process (or task) of the SoC is assigned a unique ID number. Whenever a process is called, its ID number and the time ticker are stored as an entry signature. The process ID number and time ticker are logged again at the end of the process to create an exit signature. The entry and exit signatures *do not include current reading*, in order to keep the current measurement overhead low. The resulting power consumption data log is analyzed off-line to determine the power consumption of each process. This is performed by analyzing the process entry and exit signatures, revealing the interaction and preemption among the processes.

It is often true that the SoC has too many fast processes that are difficult to track within the time resolution of our protocol. The designer needs to define which process should be ignored based on the given time resolution. While such processes still

may interfere with the measurements, their combined effect can be "calibrated out" by observing the log of a given process along an extended path of repeated executions.

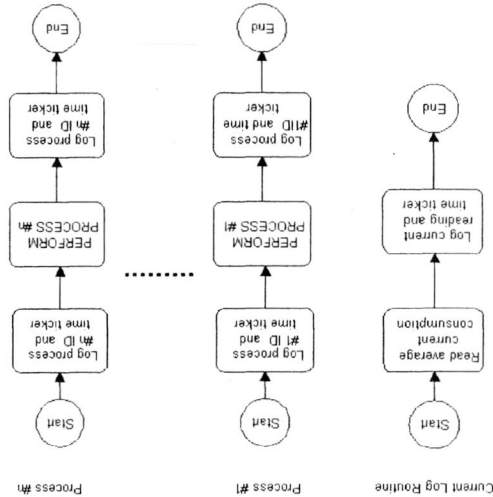

Figure 1: Data Logging of Multiple Processes

4. THE SoC DEMONSTRATION TOOL

We present a demonstration tool of an emulated SoC in which the voltage across a *single* small series resistor R_s at the power input is measured by a high-side current sense amplifier (Maxim Semiconductor MAX4172 [9]). We have emulated a typical SoC design using a common mixed signal micro-controller (C8051F321) combined with the external functional circuitry shown in Fig. 2. The design allows the user to control the access rate of the DRAM memory devices to demonstrate how the reduction in memory access rate reduces the current consumption. The entire circuitry can be readily integrated within a SoC. As mentioned, we use an A/D converter channel that features extensive averaging capabilities. Therefore, it is sufficient for the system to read the current average 10-50 times per second in order to create a detailed real-time profile of the current consumption.

Our SoC emulator board is interfaced to a custom graphic program running on a PC. The PC obtains a log of the instantaneous overall power consumption of the SoC in order to demonstrate the clear correlation between the active process and the measured power consumption. Our tool can demonstrate how current consumption is affected in real-time when the SoC performs different internal operations like DRAM accesses, SRAM accesses, or LED (fixed current consumption) activities. The user can setup the test flow for the desired mixture of activities to be performed at different durations.

The dynamic power consumption of P_{Total} in equation (1) is demonstrated during the DRAM access when the user changes the SoC frequency. The LED loads provide an example of a process with fixed power consumption.

While this SoC emulator uses a single voltage power source, modern SoC designs call for multiple voltage sources as well as voltage scaling for power reduction. Our approach can still work in such systems with one measurement point at the main power entry to the SoC.

It is important to note that our approach is intended to be implemented within the framework of a power consumption refining paradigm without real-time PC interface. The SoC simply keeps a log of the continuous current measurements together with the corresponding ID markers for off-line analysis.

5. EXPERIMENTAL RESULTS

Since standard benchmarks for our approach are not yet established, we have captured the current consumption logs shown in Fig. 3 while running radically different test flow settings. In Fig. 3.a we switch various resistive and memory loads that run for relatively long periods. This setup produces almost step function changes in the overall instantaneous current consumption. In Fig. 3.b we have reduced the duration of the test periods to illustrate how our approach provides continuously report of the power consumption. Fig. 3.c illustrates the lower DRAM power consumption when using reduced board frequency.

Table 1 shows the typical current consumption obtained with our tool using different process mixtures. In the first line we have listed the base-line current which is obtained when turning off all the processes. The 35mA reading is therefore the current consumption overhead of the SoC effort in communicating with the PC, running the current measurement process, and performing other "standby" functions. This value is subtracted from all the subsequent readings of Table 1 to obtain the net process currents. This second row shows the 22mA net current reading for the LED (fixed load) obtained by subtracting the base line current of 35mA from the 57mA overall reading. Similarly we obtain the net process current in rows 3 and 4 for the DRAM only (33mA) and SRAM only (16mA) processes. In line 5 we calculate the net current for the DRAM + SRAM + LED process to be 75mA. This reading slightly deviates (6%) from the expected sum of 71mA when the processes run separately (lines 2-4).

Table 2 summarizes the performance of our development system using the test flows of Fig. 3. In addition to the maximum and minimum currents taken from the scrolling power consumption graph, we analyzed the performance with various polling rates. The slowest polling rate is defined as the rate at which the SoC measurements start to average out the current consumption differences among the processes. The fastest polling rate (2mS) indicates the limit of our demonstration system, taking into account the SoC emulator's computing power and the limitation on the serial channel interface.

Fig. 3 and Table 2 illustrate the viability of our approach to correlate power consumption data to the actual processes at a relatively high resolution. Even the low overhead of 30mS polling allows a detailed analysis for all the three different flows in Fig. 3. Since the PC was a fast 3GHz dual Xeon computer, the overhead at the PC side is negligible for the current tests.

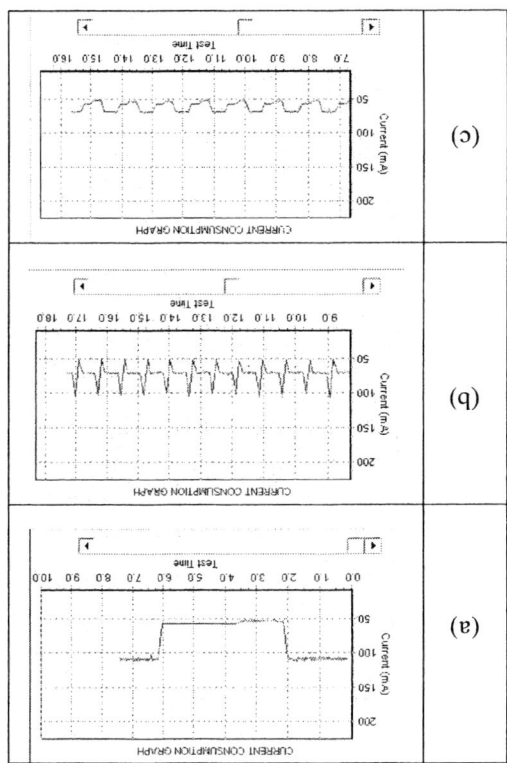

Figure 3: Running Different Test Flows

Table 1: Computing the Current Consumption of the Individual Processes

Process mixture setting	Overall current consumption	Net process current
SoC base-line	35 mA	--
LED only	57 mA	22 mA
DRAM only	88 mA	33 mA
SRAM only	51 mA	16 mA
DRAM + SRAM + LED	110 mA	75 mA

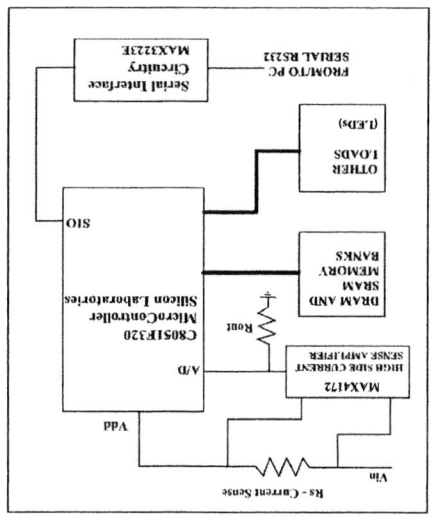

Figure 2: A SoC emulator utilizing our Power Measurement Technique

6. CONCLUSIONS

We suggest a simple enhancement for the SoC design that enables the designer to obtain a detailed insight into the power consumption of each component and related process. Such detailed information can be used to fine tune and validate the power simulation tools within the design framework. The detailed power consumption data can be advantageously used in the power design refining process, identifying those hardware/software components of the SoC design that need special attention to meet the design power budget.

We have developed an efficient logging algorithm that achieves detailed power consumption measurements for each process using a single current sensor. The sensor measures the instantaneous current consumption at the power entry point to the SoC. It requires minimal resources that are abundantly available in the SoC with minimal cost implication.

We have demonstrated our approach with a development tool that emulates a SoC in conjunction with a PC interface. Using the internal averaging feature of the typical A/D channels of the SoC, the execution time overhead for the detailed power consumption measurements was minimal.

Our approach may be implemented in a DPM environment to provide real-time power consumption data to better steer the adaptive power control algorithms.

References

[1] L. Benini, A. Bogliolo, and G. De Micheli, "A Survey of Design Techniques for System-Level Dynamic Power Management", *IEEE Trans. VLSI Systems*, vol. 8, no. 3 (June 2000): 299-316.

[2] R. A. Bergamaschi, Y. W. Jiang, "State-based power analysis for systems-on-chip", In *DAC '03*, June 2003, pp. 638-641.

[3] A.P. Chandrakasan, S. Sheng, and R.W. Brodersen, "Low Power CMOS Design", *IEEE J. Solid-State Circuits*, Vol. 27, No. 4, April 1992, pp. 473-483.

[4] W. Fornaciari, P. Gubian, D. Sciuto, and C. Silvano, "Power Estimation of Embedded Systems: A Hardware/Software Codesign Approach", *IEEE Tran. on VLSI Systems*, vol. 6, no. 6, (June 1998): pp. 266-275.

[5] D. Hillman, "Using Mobilize Power Management IP for Dynamic & Static Power Reduction in SoC at 130 nm", In *DATE '05*, March 2005, 7 pages.

[6] D. E. Lackey, P. S. Zuchowski, T. R. Bednar, D. W. Stout, S. W. Gould and J. M. Cohn, "Managing power and performance for System-on-Chip designs using Voltage Islands", In *ICCAD '02*, Nov 2002, pp. 195-202.

[7] I. Lee, H. Kim, P. Yang, S. Yoo, E.-Y. Chung, K.-M. Choi, J.-T. Kong and S.-K. Eo, "PowerViP: Soc power estimation framework at transaction level", In *ASP-DAC '06*, Jan. 2006, pp. 551-558.

[8] Y. Li and J. Henkel, "A Framework for Estimating and Minimizing Energy Dissipation of Embedded HW/SW Systems", In *DAC '98*, 1998, pp. 188-194.

[9] "MAX4172 Low-Cost, Precision, High-Side Current-Sense Amplifier", Maxim Integrated Products, Inc. 120 San Gabriel Drive, Sunnyvale, CA 94086 Tel. 408-737-7600. Available on-line: http://www.maxim-ic.com/company/

[10] V. Tiwari, S. Malik, and A. Wolfe, "Power Analysis of Embedded Software: A First Step Towards Software Power Minimization", *IEEE Tran. VLSI Systems*, vol. 2, no. 4 (Dec. 1994) pp. 437-445.

[11] R. Watanabe, M. Kondo, H. Imai, H. Nakamura and T. Nanya "Task scheduling under performance constraints for reducing the energy consumption of the GALS multi-processor SoC", In *DATE '07*, April 2007, pp. 797-802.

[12] Wayne Wolf, *Computers as Components*, Morgan Kaufmann, 2001.

The SoC design framework for detailed power consumption analysis. The automated analysis allows the designer to identify which portion of the design's power budget is consumed by each process and determines which process needs to be further refined. The data is also used to fine tune the power simulation tools in the SoC design framework.

Our technique does not severely interfere with the process mixture and order in a typical SoC application. Each process logs its entry and exit point in the log, to support multiple simultaneous processes and inter-process preemption. Table 3 further demonstrates that our power consumption data log requires minimal storage area. The entire log is downloaded to the SoC design framework for detailed power consumption analysis.

Table 3: Demonstration of Offloading Power Consumption Data

Process ID	Time Ticker	Current Consumption
...	...	...
0	100	35
0	110	40
1	112	--
0	120	73
0	130	75
2	136	--
0	140	121
0	150	129
1	160	113
1	167	--
0	170	218
0	180	240
2	185	--
0	190	180
0	200	43
...	...	...

Table 3 illustrates the power consumption data log in an SoC designed in accordance with Fig. 1 for off-line analysis. We show the data log during an arbitrary period from 100ms to 200ms and demonstrate how to resolve the power consumption of each process when multiple processes are active together or when preempting occurs. The ID number of the continuous current log routine is set to 0 and it is called every 10ms. We demonstrate two processes having ID numbers 1 and 2 (process ID#1 and 2). Process ID#1 consumes more power than process ID#2. Process ID#2 starts at time 136ms, while process ID#1 is still on. In this example both processes run together, quickly bringing their average power consumption to 129mA. At time 167ms process ID#1 terminates, leaving process ID#2 working along at its higher current consumption of 240mA. As process ID#2 ends at time 185ms, the consumption gradually (due to the effect of the filter capacitor) goes back to the 43mA base-line consumption. Each process may exhibit inherent, inter-process related, power fluctuations.

Table 2: Performance results with different test flows.

Test Flow	Max Current	Min Current	Slowest Polling	Fastest Polling
Fig. 3a	110mA	53mA	250ms	2ms
Fig. 3b	100mA	52mA	60ms	2ms
Fig. 3c	70mA	50mA	150ms	2ms

A High-Performance Multi-Match Priority Encoder for TCAM-Based Packet Classifiers

Miad Faezipour and **Mehrdad Nourani**

Center for Integrated Circuits & Systems
The University of Texas at Dallas, Richardson, TX 75083
{mxf042000,nourani}@utdallas.edu

Abstract—**This paper introduces a high-speed and low power multi-match priority encoder design applicable in many computer and networking systems. We propose a scalable multi-match prioritizer logic circuitry that can successively find all or the first r matched inputs in a set. The design is well suited for multi-match packet classification tasks that utilize content addressable memories as the search engine. We use a data partitioning scheme to efficiently reorganize input data for further performance improvement. A VLSI implementation of our design in 0.18μm technology can achieve speed that outperforms the conventional multi-match packet classifier design by more than an order of magnitude. Overall power consumption is reduced by more than 40% using innovative partitioning which limits the search to a small portion of TCAM cells.**

I. INTRODUCTION

Priority-based operations are fundamental to a variety of basic digital components such as incrementers/decrementers and comparators. From networking perspective, they are widely used in packet forwarding and packet classification as well. A Priority Encoder (PE) circuit encodes the address (index) of the highest priority input. The operation is performed by passing a priority token from the highest priority bit (input) to the lowest. Performance of PEs are critical, as they are integrated within many digital systems. As the number of inputs scales up, high-speed and low power PE circuits are essential in order to achieve high-performance systems. High-speed and power efficient CMOS PE circuits have been introduced in [1][2][3]. Multi-level folding and parallel priority look ahead techniques were the key novelties used in these designs.

In general, packet classification refers to finding the best matching filter containing multiple fields among the *filter* (also called *rule* in the literature) set for a given packet. The standard five-tuple fields include the source address, destination address, protocol, source port and destination port [4]. Ternary Content Addressable Memories (TCAMs) are well suited for performing high speed parallel searches on database with ternary entries, since they provide the match results with deterministic throughput (i.e. one search per cycle) and deterministic capacity. TCAMs inherently include the priority encoder, resulting in the highest priority match. Hence, TCAM has become quite popular for packet classification [4][5].

Multi-priority encoding is used in a number of new emerging networking systems. One main application of multi-priority encoding is *multi-match packet classification* using Ternary CAM devices. Network Intrusion Detection Systems (NIDS), load balancers, and programmable network elements (PNE) require finding all or the first few matching filters in packet classification. Regev et al. introduced logic circuit designs that utilize multiple single priority encoders to find matching addresses successively [7][8]. Nowadays, gigabit rates are required for networking applications. Therefore, developing high-speed and low-power multi-priority encoder circuits are design concerns as well. While TCAMs perform packet classification at high speed, they cannot directly report all possible matches in a database. This is due to the native structure of a TCAM cell design, which consists of a conventional priority encoder, resulting in the highest priority match. We propose a multi-matching packet classifier by modifying the prioritizer circuit in conventional TCAM units.

II. MULTI-MATCH PRIORITY ENCODING

A TCAM cell includes the TCAM word and a PE unit. The PE unit itself consists of a prioritizer circuit and a conventional address encoder. A valid log_2n-bit address out of the encoder is generated only if at most one of its n inputs is high at a time. The main idea is to modify the single-match prioritizer unit to a Multi-match Prioritizer (MPZ), as shown in Figure 1, so that the encoder would generate all matching indices, one at a time.

An 8-bit power optimized priority encoder cell is used as our reference model for the single-match prioritizer (PZ) unit [1]. We have designed and implemented the transistor level schematic of this prioritizer circuit shown in Figure 2. The logic equation for the PZ circuit (using boolean notations) is:

$$EP_i = en \cdot \left(\prod_{k=0}^{i-1} \overline{D_k}\right) \cdot D_i, \qquad 0 \leq i \leq n-1 \quad (1)$$

Equation 1 indicates that the PZ circuit has n inputs and n outputs where EP_i denotes the ith output, D_i's are the input lines and en is the enable line. In the PZ circuit, only one chain of the evaluate transistors will discharge during evaluation. The transistor count of the circuit is 62, and the necessary precharge nodes are very few. We use this circuit as the basic (single) prioritizer unit. It is also highly power efficient compared to other priority encoder configurations using multi-level look-ahead structures [2][3].

A. Multi-Match Prioritizer Unit

We add a control logic circuitry to the prioritizer circuit to report all matches in a prioritized sequence. The MPZ

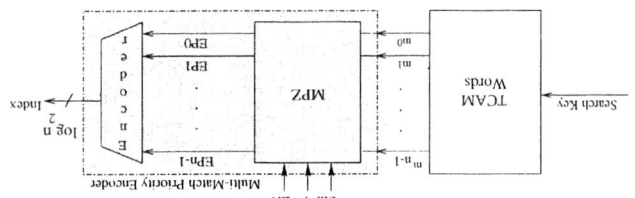

Figure 1. Conceptual block diagram of the design.

circuit, shown in Figure 3, functions in response to a counter which counts from 0 to n, where n is the highest possible number of matches. In other words, n can be assumed as the number of inputs in the worst case. On the first clock cycle, the MPZ should function as a single-prioritizer unit, reporting the highest priority match. On the next clock cycle the next highest priority match should be provided at the output. This procedure should be followed for all other clock cycles until the counter has reached counting up to n. In each clock cycle, a function of the original inputs and the higher priority outputs of the prioritizer circuit in the previous clock cycle should be fed through the prioritizer circuit. Let m_i denote the original input lines (i.e. match lines from TCAM words), ep_i denote the EP_i outputs of the prioritizer after one clock cycle. M_i be the set of inputs that should be given to the prioritizer circuit, and EN be the enable line. The logic equations for the MPZ circuit can be derived as:

$$M_i = \bar{s} \cdot m_i + s \cdot \sum_{k=0}^{i-1} ep_k, \qquad 0 \le i \le n-1 \qquad (2)$$

$$EP_i = EN \cdot \prod_{k=0}^{i-1} \overline{M_k} \cdot M_i, \qquad 0 \le i \le n-1 \qquad (3)$$

Signal s is the select line of the multiplexers that control which data should be chosen for the corresponding M_i. This select line should be low for the first clock cycle and high for the rest. Thus, s can be implemented by simply ORing all the counter outputs c_i: $s = \sum_{k=0}^{\log_2 n - 1} c_k$. The MPZ unit functions in an efficient manner; in essence it reports all r matches in exactly r-cycles. This implies that in case of the need to report the first r matches instead of all possible matches, a comparator unit could be added to the MPZ design wherein the count value c and the r value are compared. Once the count exceeds r, the enable line EN is set to zero; hence disabling the MPZ unit.

As authors in [9] stated, the maximum degree of matches (number of matches) often requested in real-world access

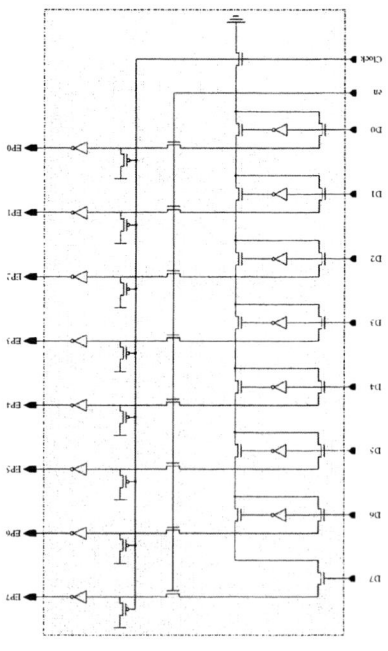

Figure 2. The reference 8-bit power-optimized prioritizer circuit.

control list (ACL) filter database is statistically around 8. Considering this fact the counter used in any MPZ unit can be designed to count up until 8. This implies that only $log_2 8 = 3$ lines are required for the counter. In addition, the comparator can be designed to compare the count value and $r = 8$.

B. Scalability

To achieve a modular design with cascaded blocks we define an *output enable* (OE) line which indicates when all the matches have been provided at the output. This signal is activated when all the matching results have been provided at the output, and deactivated when any match is found at the output. The OE signal also highly depends on the EN line. The OE line in a MPZ can be expressed as follows:

$$OE = EN \cdot \prod_{i=0}^{n-1} \overline{EP_i} \qquad (4)$$

Figure 4 shows the concept of cascading eight, 8-bit MPZ modules to design a 64-bit MPZ. By connecting the OE line of each stage to the EN (enable line) of the next stage (in this case higher priority stages are placed at the left), we assure that each block would be enabled only if all the higher priority blocks have completed reporting their matches at the output. The cascaded design would have at most two additional clock cycle delays for each mismatching MPZ unit. Hence, if all matches are concentrated within one block of MPZ, a high throughput can be achieved.

Another scalable design for the MPZ is based on the concept of parallelism. A parallel architecture similar to the parallel priority look-ahead technique [1] is shown in Figure 5. In case of 64-bit multi-match design, the first stage MPZ unit should have a clock period of at least 8 times slower than the 8 MPZ units in the second stage. This is due to the fact the first stage should be enabled for at least 8-cycles, to allow the second stage MPZ to report all the results.

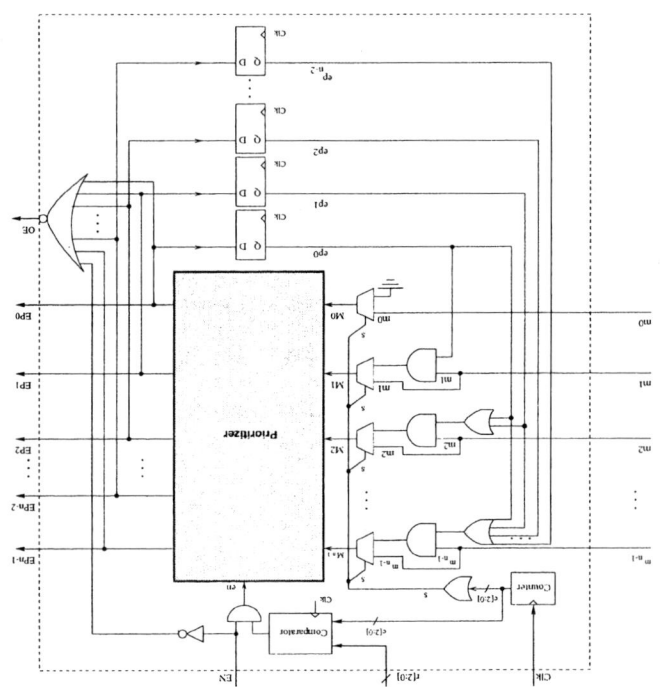

Figure 3. The MPZ architectural design

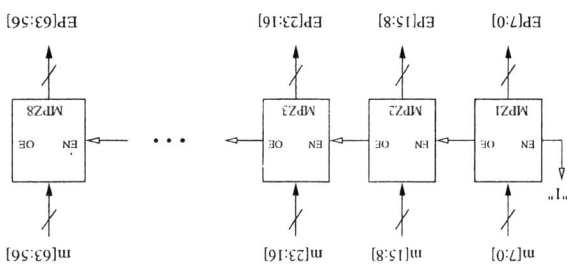

Figure 4. Cascaded MPZ architecture for a 64-bit design.

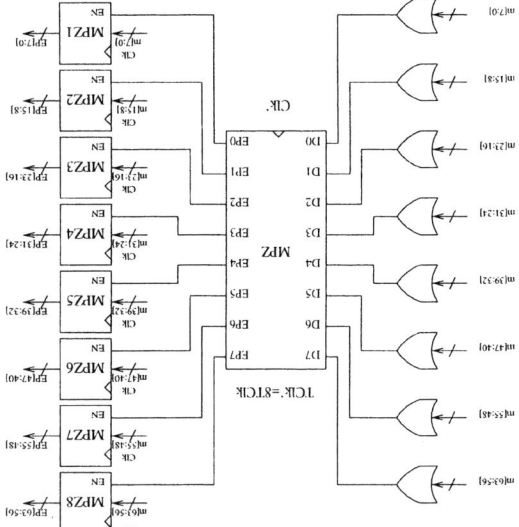

Figure 5. Parallel MPZ architecture for a 64-bit design.

The parallel architecture has a better speed for single-match applications (when assuming $T_{Clk} = T_{Clk}$) and multi-matches concentrated in one MPZ unit, while the cascaded configuration is observed to be more efficient in terms of both speed and area for uniformly distributed multi-matching tasks. Area is another concern for the two scalable configurations. The parallel architecture would have an additional MPZ unit plus a few OR gates, compared to the cascaded architecture.

The above discussion indicates that performance in large filter sets highly depends on the locations of the matches. To achieve best timing results, the input data filters should be relocated intelligently. This issue follows next.

III. PERFORMANCE/POWER IMPROVEMENT

Intersection among filters results mainly in the database in multiple matches. Therefore, partitioning the filter set based on filter intersections, and performing the TCAM search based on a partition, can significantly improve the performance. From the MPZ design, we have seen that by having all matches concentrated within one block, the multi-match performance can achieve the highest extreme in performance; in other words it is capable of finding r matches in r cycles.

The data partitioning scheme partitions the filters in the database such that each partition would hold the maximum number of intersections among its filters. In addition, partitions will be disjoint, i.e. any pair of partitions do not have any overlap in the filters that they contain. Since there would always be a number of filters that do not have any intersection with any other filter, one last partition is needed in which all these distinct filters can be placed.

We define the term *distance* for any two filters $f_i[w-1:0]$ and $f_j[w-1:0]$ of w-bits wide to be the summation of the bit-wise comparisons. In this definition, a don't care in any bit position would result in zero distance for that particular bit position. So, for example filters 01101x1x and 01x0xx11 have zero distance, since the result of comparison in each bit position yield zero distance. However, filters 01101x0x and 01x0xx11 do not have zero distance, since their values differ in bit position $w = 1$.

Our partitioning heuristic is to place all zero distance filters within one partition. Figure 6 (a) illustrates an example of a small filter set of 10 filters partitioned based on maximum intersections. Filters are assumed to be 8-bits long for simplicity. Filters 1, 4, and 10 have zero distance, hence they can form one partition, P_1. Also, filters 8 and 9 have zero distance with filter 10, therefore they are also placed in P_1. Filters 5 and 6 make a zero distance with filter 2, and form partition P_2. Finally, filters 3 and 7 that have no zero distance with any other filters, form the distinct filter collection, and are placed in a separate partition (P_3). In this example, if "11010010" arrives as the search key, partition P_1 would be chosen and the search is performed on a relatively small TCAM containing partition P_1 only.

Our data partitioning would ensure that all possible matches for a given packet are located in one partition. Figure 6 (b) illustrates the partitioning approach used with the MPZ architecture. Note that the last partition (P_{N_p}) does not need a MPZ unit since it would result in at most one match.

The MPZ and encoder circuit connected to the TCAM provide the addresses of the r matches in at most r cycles successively. Since the maximum number of multiple matches is around 8 [9], no larger than 8-bit MPZ units is required. Additionally, this approach does not add any extra filter to the whole set, unlike others, e.g. the SSA scheme [10], that adds new filters for partially overlapping filters in each partition. This feature makes the our approach much more memory efficient. Moreover, performing the TCAM search on only a small portion of the entire database can significantly save power consumption due to the frequently charging and discharging of the highly capacitive match line. To achieve such power savings, a *Selector Unit* is required so that the search mechanism would result in enabling the TCAM search on one partition, while disabling others. The selector unit is a small TCAM unit that stores a common code for the filters of each partition. More description on the selector design can be found in [11].

IV. EXPERIMENTAL RESULTS

A. Simulation of MPZ Unit

An 8-bit MPZ unit was implemented using Synopsys tools [12]. Timing simulations are shown in Figure 7. The MPZ inputs $m[7:0] = $ "01011000" are assumed to be the TCAM output match lines. Output results $EP[7:0]$ were observed to be "00001000," (08H) "00010000," (10H) and "01000000" (40H) on the first three clock cycles, respectively. Figure

(a) Partitioning

Before Partitioning

Filters:
1) 1101xx10
2) 0xx01110
3) 10110101x
4) 11xxxxx1
5) x0101000
6) 01101x10
7) 0111x010
8) 100x1xx0
9) 1001xx10

After Partitioning

Partition P1:
1) 1101xx10
4) 11xxxxx1
6) 01101x10
9) 1001xx10

Partition P2:
2) 0xx01110
5) x0101000
8) 100x1xx0

Partition P3:
3) 10110101x
7) 0111x010

(b) The classifier engine

Search Key → Selector →
TCAM Partition P1 (F1, F4, F10, R, F9) → MPZ + Encoder → Index
TCAM Partition P2 (F2, F5, F16) → MPZ + Encoder → Index
TCAM Partition Pn (F3, F7) → MPZ + Encoder → Index

Figure 6. Application of data partitioning to a small example.

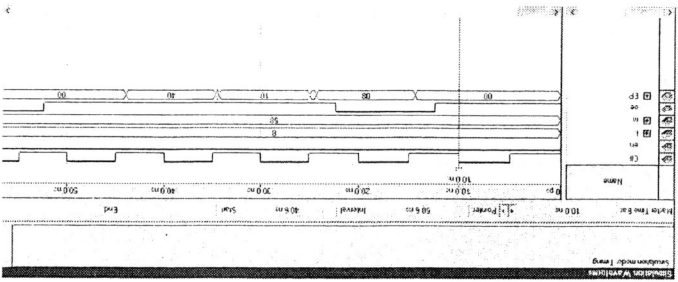

Figure 7. Timing simulation for an 8-bit MPZ unit

B. Speedup

7 indicates that the MPZ unit can work with $T_{clk} = 10ns$ ($f_{clk} = 100MHz$). Assuming the minimum length of 40 bytes per packet, MPZ can achieve throughput of $40 \times 8 \times 10^8 =$ 32Gbps.

The maximum throughput of a single MPZ unit would be the smallest clock period for the circuit to function correctly. Table I compares the statistics for four n-bit MPZ units we have implemented. As n grows, the delay and the cost of MPZ both increases. However, note carefully that growth of critical path delay, shown in the second column, is not an indication of improvement. To be more clear about this the third column shows the estimated speedup (S_n) in each case compared to a software-based classification approach. We assumed that the host running the classification software needs at least one memory access and one comparison instructions per entry. This yields speeds up of: $S_n \geq \frac{r \cdot n \cdot (t_{mem} + t_{cmp})}{r \cdot T_{clk,r}}$, where t_{mem} and t_{cmp} refer to the corresponding instructions, and $T_{clk,r}$ is the duration of the clock driving a n-bit MPZ. Obviously, $T_{clk,r}$ should be larger than the delay values given in Table I. Note that in practical cases $r << n$ and thus as n grows the speedup will be even larger than 40-73 given in the table because in our approach the overall performance depends on r and not n. Such property makes our circuit quite attractive in applications that require fast multi-match classification.

The speed of our architecture is also higher than conventional hardware based designs. A conventional TCAM delay can be written as $T_{TCAM} = T_{word} + T_{PE}$, where T_{word} is the delay of the TCAM word and T_{PE} is the delay of the priority encoder. For large TCAMs $T_{word} \approx T_{PE} \approx T_{TCAM}/2$. Our priority encoder (the MPZ unit along with an encoder) approximately has a delay of $1.4 \times T_{PE}$. A conventional multi-match TCAM based approach would spend $r \cdot (T \times T_{TCAM})$ cycles for finding r matches [9]. Hence, we obtain speedup of $S = \frac{T_{TCAM}/2 + 1.4 \times T_{TCAM}/2}{r \cdot (T \times T_{TCAM})}$ for finding r matches. For a maximum of 8 matches we achieve speedup of 9.18.

C. Power Estimates

Overall power consumption per search depends on the size of the partition that the selector unit triggers. Assuming $|F|$ to be the size of the entire filter set, and $|P|_{max}$ ($|P|_{min}$) to be the maximum (minimum) size of a partition, the maximum power savings ΔE follows this formulation: $\frac{|F|}{|P|_{max}} \leq \Delta E \leq \frac{|F|}{|P|_{min}}$. Table II shows the maximum and minimum power savings achieved by partitioning based on the data obtained from filter sets that had 5000 filters of bit-width 150 each, which is typical for real databases. Clearly, significant power reduction is achieved.

TABLE I
SIMULATION RESULTS FOR n-BIT MPZ UNITS

n	Delay [ns]	Area [NAND]	Speedup S_n
8	9.92	206	40.32
16	17.92	437	44.64
32	25.09	769	63.77
64	43.87	1373	72.94

TABLE II
POWER SAVINGS

$|P|_{max}$	$|P|_{min}$	ΔE_{max}	ΔE_{min}
4245	12	99.76%	15.1%
4225	8	99.84%	15.5%
4090	4	99.92%	18.2%

REFERENCES

[1] C. Kun, S. Quan and A. Mason, "A Power Optimized 64-bit Priority Encoder Utilizing Parallel Priority Look-Ahead," IEEE ISCAS'04, vol. 2, pp. 753-756, May 2004.

[2] C. H. Huang, J. S. Wang and Y. C. Huang, "Design of High-Performance CMOS Priority Encoders and Incrementer/Decrementers Using Multilevel Lookahead and Multilevel Folding Techniques," IEEE Journal of Solid-State Circuits, vol. 37, no. 1, pp. 63-76, Jan. 2002.

[3] J.S. Wang and C.H. Huang, "High-Speed and Low-Power CMOS Priority Encoders," IEEE Journal of Solid-State Circuits, vol. 35, no. 10, Oct. 2000.

[4] K. Zheng, H. Che, Z. Wang, and B. Liu, "TCAM-Based Distributed Parallel Packet Classification Algorithm with Range-Matching Solution," INFOCOM'05, 2005.

[5] E. Spitznagel, D. Taylor and J. Turner, "Packet Classification Using Extended TCAMs," in Proceedings of the 11th IEEE International Conference on Network Protocols, pp. 120-131, Nov. 2003.

[6] F. Yu, R. H. Katz and T. V. Lakshman, "Efficient Multimatch Packet Classification and Lookup with TCAM," in Proceedings of the 12th Annual IEEE Symposium on High Performance Interconnects HOTI'04, pp. 28-34, Aug. 2004.

[7] Z. Regev, "Multi-Priority Encoder," US Patent 20040125808, July 2004.

[8] A. Regev and Z. Regev, "Priority Encoder for Successive Encoding of Multiple Matches in a CAM," US Patent 20040125633, July 2004.

[9] K. Lakshminarayanan, A. Rangarajan and S. Venkatachary, "Algorithms for Advanced Packet Classification with Ternary CAMs," ACM SIGCOMM'05, Aug. 2005.

[10] F. Yu, T. V. Lakshman, M. A. Motoyama and R. Katz, "SSA: A Power and Memory Efficient Scheme to Multi-Match Packet Classification," ACM Proceedings of the 2005 Symposium on Architecture for Networking and Communications Systems ANCS '05, pp. 105-113, Oct. 2005.

[11] M. Faezipour, "High-Speed Multi-Match Packet Classification Using TCAM," Masters Thesis, UTD-EE-11-2006, Nov. 2006.

[12] Synopsys Inc., "User Manuals for SYNOPSYS Toolset Version 2005.06," 2005.

An Efficient Implementation of Scalable Architecture for Discrete Wavelet Transform On FPGA

Michael GUARISCO, Xun ZHANG, Hassan RABAH and Serge WEBER

Nancy University

Laboratoire d'Instrumentation Electronique de Nancy (LIEN)

Vandoeuvre-les-nancy, Nancy, 54500

email: {michael.guarisco, xun.zhang, hassan.rabah, serge.weber}@lien.uhp-nancy.fr

Abstract—This paper presents efficient reconfigurable architecture to perform discret wavelet transform. This architecture, which is based on FPGA technology, consists of a reconfigurable processing module, reconfigurable controller, data organization unit and adresse generator, and on chip memory. The reconfigurable adresse generator and controller handles a flexible and efficient address generation for an efficient data memory access and bandwidth. This architecture is scalable and allows processing of a continuous data flow in real time and for any number of levels. The practical working of the architecture is explained and its hardware implementation on Xilinx Virtex-5 FPGA is reported.

I. INTRODUCTION

The discrete wavelet transform (DWT) allows to separate, on one or more levels, the low frequencies and high frequencies components of a signal, or, in the event, of an image. It is a very powerful tool and is a good alternate to the discrete cosinus transform (DCT). The DWT is used in the new compression standard JPEG2000 [1]. and was the subject of numerous research and many different architectures [2]. have been built. Some of these are implemented using pipelining but they are highly dependant of the number of levels to compute [4], [5]. This presents a disadvantage since if we need to calculate more levels at a given time of the treatment, then, it will be necessary to modify architecture. Other architectures use the same hardware for each level, but, due to the need to re-use the data resulting from calculation, these architectures are not appropriate to the real-time computation. In this paper, the main goal is to maximize the effectiveness (in term of speed and cost in memory access) of the computation and minimize the necessary hardware ressources. We have searched to work out a scalable architecture which can perform DWT on the basis of any filter type. In the first part of this paper, we shall briefly describe the wavelet transform process. We will then describe the working principle of our architecture. And we will finish by giving some results of simulation.

II. THE DISCRETE WAVELET TRANSFORM

The two dimensional (2D) discrete wavelet transform (DWT) is a rapid decomposition in the multimedia application domain. The DWT is computed by successive low-pass and high-pass filtering (figure 1). The output of each filter is decimated that is every second value is removed halving de length of the output [6]. The output of each filter stage is made of transform coefficients and each filter stage represents a level of transform. The low pass result is then transformed by the same process and this is repeated until the desired level is reached.

A. Classical processing approach

The classical approach to 2D transform is to process each layer in the tree separately and to process the row and column layers successively one after the other. The performance of this approach is strongly limited by the management of temporary data required between two successive layers and between horizontal and vertical filtering. For a 2D image with N rows and N columns and L levels, the amount of data to be filtered on each layer increases by a factor of four from one layer to the next, and the total amount of processed data along the whole tree decomposition process is given by the following equation:

$$D = \sum_{t=1}^{L} \frac{4^{t-1}}{4^{L-1}} \times N \times N = \frac{4^{L}-1}{3 \times 4^{L-1}} \times N \times N \quad (1)$$

To process a $N \times N$ image, a temporal memory of size

$$D - N \times N = (1 - \frac{4^{L}-1}{3 \times 4^{L-1}}) \times N \times N \quad (2)$$

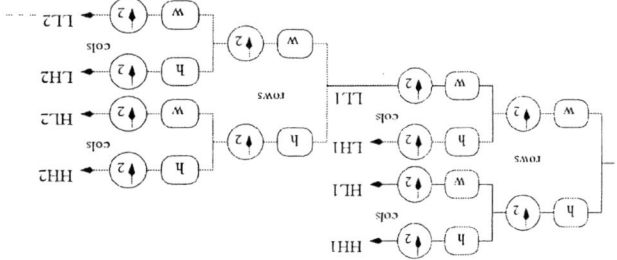

Fig. 1. 2-D Discret wavelet transform

is required. As an example, for 2 level resolution a temporal memory of $0.25\ N \times N$ size is required. For a given layer, the filtering process is achieved horizontally and vertically; thus two read accesses and two writes accesses are necessary and the total amount of data read and written is expressed as $D_w = D_r = 2 \times D$. The memory bandwidth B, in bidirectional access case, can be considered as the product of the total amount of data processed for a frame per second (fps) $T_{df} = (D_r + D_w) \times fps$ and the number of bits N_b of a coefficient:

$$B = T_{df} \times N_b \qquad (3)$$

As an example, for a gray level image of 512×512 pixels with 25 frame per second, 8 bits per pixel and 2 levels of reconstruction, a bandwidth of 260 Mb/s is required. These results illustrate the memory management problem as the main bottleneck of the classical approach.

III. PROPOSED PROCESSING APPROACH

In order to reduce the memory size and to optimize the over-all system performance, the wavelet algorithm is redesigned to exploit efficiently the inherent processing parallelism. This processing parallelism is possible if the required data is accessible in parallel, accordingly a data partitioning is used. The proposed organization is based on the use of fragmented double port on chip memory allowing parallel and simultaneous read write accesses. Moreover, the same memory is used to store intermediate computation results.

A. System Architecture

The system architecture is depicted in figure 2. It is composed of three on chip blocs of memory, each one of the size of a picture or the size of a macroblock. While a memory bloc is used for computation of the DWT, a second holds the entering data, and a third is needed to reconstitute a data flow which represents the transformed image. In addition to these three blocs, the system is composed of a reconfigurable processing unit two data organization units and control unit.

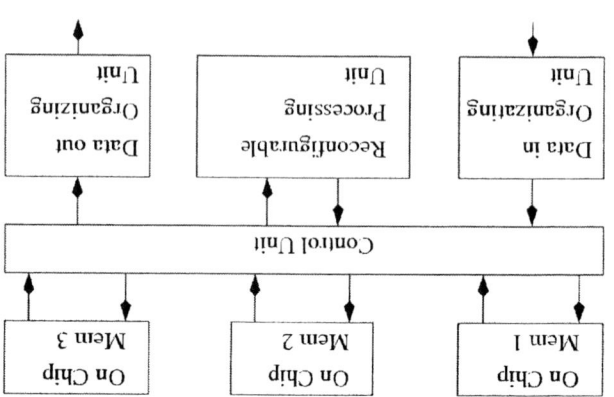

Fig. 2. System Architecture

1) *Reconfigurable Processing Module:* The reconfigurable processing module corresponds to the association of the processing unit and a memory bloc as shown in figure 3. Each of the three memory blocs are divided in four separated dual port memory blocs. The controller can reach each bloc at same time. The reconfigurable processing unit is the more important part and the one which takes many places inside the reconfigurable logic. It contains state machine to control the execution of the DWT by the two processing elements (PE1 & PE22). Each processing is composed of two reconfigurable data-paths implementing two filters (one for the high frequencies coefficient, the other for the low frequencies coefficient). The filters used in our architecture are the 5/3 and 9/7 [3], [7].

2) *Data Organizing Unit:* The system uses two data organisation modules. The first organizing module dispatches the incoming image data into the different fragments of memory bloc to allow the processing parallelism. The second organizing module reconstruct the order of the computed DWT coefficients.

3) *Control Unit:* One of the key elements on the architecture is its controller unit which connects the right memory to the right Unit at the right time. In fact, for example once the memory bloc $Mem1$ is full (the memory bloc $Mem3$ is empty in the same time and the DWT is computed in the memory bloc $Mem2$), each memory bloc is translated to connect with the adjacent module. The empty memory is then used to store new data, the full memory which contains not yet treated data is connected to the processing unit and the memory bloc which contains the treated data is connected to to the data out organizing module to be emptied.

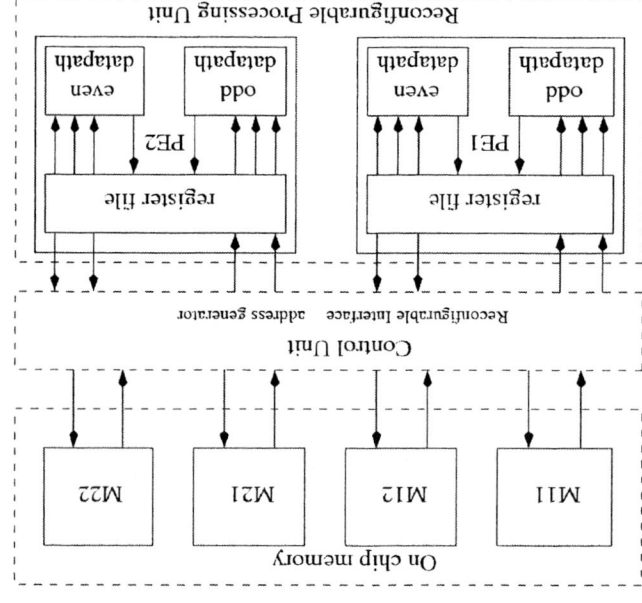

Fig. 3. Reconfigurable Processing Module

B. Detailed operations

To explain the operating details of the system, consider an original 8x8 image as shown in figure 4.a. The role of the Data in organizing unit is to rearrange the the pixels in the memory bloc. In order to benefit from the parallelism, it arranges the pixels as shown in figure 4.a. So, due to the utilisation of pixels divided in four independent dual port memories, the processing controller can reach, for a given i, S_i and D_i, which are normally two consecutive pixel in the image and those which we need to calculate at the same time the two coefficient of the DWT. If we want to calculate two new samples at each clock cycle, we have to reach two consecutive elements (S_i and D_i) at the same cycle.

So, in a first time, each processing element can calculate 1D-DWT in line. As we have two element, the system can compute two 1D-DWT in the same time. In a second time, the system computes the 1D-DWT, but now, in columns.

Thus, we save a precious time and we can theoretically achieve an infinite number of levels. Let be T_{load} the needed time to fill a memory bloc at the frequency of the data (it corresponds to the time of a complete reading or writing, pixel after pixel, of the whole memory bloc), we can say that the execution time of the first level is $0.5 * T_{load}$ (we reach two pixels at one clock cycle, so, it divides by two T_{load}, we have two PE that divides again by two T_{load}) but we have to achieve two time the 1D-DWT. Finally the execution time is of $0.5 * T_{load}$. Moreover, we know that for the next level, we need only the low frequency coefficients which represents only a quarter of the total result of the previous level. The execution time is then the result of an arithmetic suite which is represented in equation 4.

$$T_{exec}^n = \sum_{i=1} \frac{1}{4^{i-1}} \frac{T_{load}}{2} \qquad (4)$$

If the number of level tends towards infinite, the execution time is then of $2/3 * T_{load}$. The Data out Organizing unit allows getting back the DWT coefficients in an ordered way. This controller can be easily modified to adapt the structure of the data flow of the system.

a: Original Image

S00	S01	S02	S03	D00	D01	D02	D03
S04	S05	S06	S07	D04	D05	D06	D07
S08	S09	S10	S11	D08	D09	D10	D11
S12	S13	S14	S15	D12	D13	D14	D15
S16	S17	S18	S19	D16	D17	D18	D19
S20	S21	S22	S23	D20	D21	D22	D23
S24	S25	S26	S27	D24	D25	D26	D27
S28	S29	S30	S31	D28	D29	D30	D31

b: Reorganized Image

S00	D00	S01	D01	S02	D02	S03	D03
S04	D04	S05	D05	S06	D06	S07	D07
S08	D08	S09	D09	S10	D10	S11	D11
S12	D12	S13	D13	S14	D14	S15	D15
S16	D16	S17	D17	S18	D18	S19	D19
S20	D20	S21	D21	S22	D22	S23	D23
S24	D24	S25	D25	S26	D26	S27	D27
S28	D28	S29	D29	S30	D30	S31	D31

Fig. 4.   Memory organization

IV. SIMULATION RESULTS

We have modeled the architecture in HDL in the sofware suite ISE from Xilinx. The simulation results agree with our theoretical waiting. Indeed, we can perform with this architecture a very high number of levels. According to the simulation results, we can run on a working frequency of 67MHz. But as we use the internal memory of an FPGA, we are limited and we can reach an image size of only 128x128 pixels. The solution consists of a small modification of the data organizing units to allow the architecture to treat macro-bloc instead of a whole picture.

V. CONCLUSION AND FUTURE WORK

In this paper, we have described a scalable and reconfigurable architecture for an efficient DWT implementation for multimedia applications. This architecture can be used for any types of filters, any size of image and any level of transformation. The memory is organized as a set of independent memory blocs. Each memory bloc is a reconfigurable module. The scalability of the architecture is achieved through the flexibility of memory blocs and processing elements to match the desired resolution. The on-chip memory is used not only to hold the source image, but also to store the temporary and final result. Hence, there is no need of temporal memory. The processor has no instructions and then no decoder; in fact, the hardware reconfigurable controller plays the role of a specific set of instructions and their sequencing. For a given set of tasks, a set of configurations are generated at compile time and loaded in run time by the configuration manager via configuration memory.

REFERENCES

[1] "JPEG2000 Image Coding System", JPEG 2000 final committee draft version 1.0, March 2000 (available from http://www.jpeg.org/FCD15444-1.htm).

[2] C. S. Burrus, R. A. Gopinath and H.Guo. "Introduction to Wavelets and Wavelet Transforms - A Primer", Prentice Hall, New Jersey, USA, 1998.

[3] G.Strang and T.Nguyen, "Wavelets and Filter Banks", Wellesley-Cambridge Press, 1996.

[4] S.Masud "VLSI system for discrete wavelet transforms", PhD Thesis, Dept. of electrical engineering, The Queens University of Belfast, 1999.

[5] F.Marino, "A Double-Face' Bit-Serial Architecture for 1-D Discrete Wavelet Transform" IEEE Trans. Circuit Syst.II, Vol. 47, NO.1, pp 65-71, Jan 2000.

[6] S. Mallat.A Theory for Multiresolution Signal Decomposition: The Wavelet Representation, IEEE Transactions on Pattern Analysis and Machine Intelligence, Vol. 11, no. 7, pp674-693, July 1989.

[7] M.Nibouche, A.Bouridane and O.Nibouche Rapid Prototyping Of Biorthogonal Discrete Wavelet Transforms on FPGAs To appear in IEEE International Conference on Electronics, Circuits and Systems, Malta, Sep 2001.

A New Fast Slew Buffering Algorithm Without Input Slew Assumptions

Shiyan Hu, Jiang Hu

Department of Electrical and Computer Engineering
Texas A&M University
College Station, Texas 77843

Abstract—As VLSI technology moves to the nanoscale regime, an ultra-fast slew buffering technique to buffer large number of nets and minimize buffering cost is highly desirable. The existing method proposed in [1] is able to efficiently perform buffer insertion with a simplified assumption on buffer input slew, however, when handling more general cases without input slew assumptions, it becomes slow despite that significant amount of buffer area savings can be obtained. In this paper, a fast buffering technique is proposed to handle this difficult general problem. Instead of building solutions from scratch, the new approach performs efficient optimizations to buffering solutions obtained with the fixed input slew assumption. Experiments on industrial netlists demonstrate that our algorithm is very effective and highly efficient. Compared to the commonly-used van Ginneken style buffering, up to 49× speed up is obtained and often 10% buffer area is saved. Compared to the algorithm without input slew assumption proposed in [1], up to 37× speedup can be obtained with slight sacrifice in solution quality.

Keywords: Buffer Insertion, Slew Constraint, Non-Fixed Input Slew, Efficiency, Physical Design

I. INTRODUCTION

As VLSI technology moves to the nanoscale regime, devices scale much faster than interconnects. As an effective timing optimization technique for interconnects, buffer insertion has been widely used in industry. For example, [2] shows that about 25% gates are buffers in two recent IBM ASIC designs. On the other hand, interconnect resistivity causes signal integrity to significantly degrade. This issue severely aggravates with advancing technologies. As such, buffers need to be inserted also for meeting slew constraints [1].

In practice, slew constraint is significantly more prevalent than timing constraint [1]. Once nets are buffered for satisfying slew constraints, most of them will automatically satisfy slew constraints. In fact, it is reported in [2] that in IBM ASIC designs, for about 95% nets, buffering based on slew is sufficient to meet their timing constraints, while only about 5% nets need to be re-buffered for timing optimization.

This suggests a better way of using buffer insertion techniques in the physical design flow [2]. Suppose that we are to buffer millions of nets. They are first buffered using slew driven buffering techniques. After performing timing analysis on the resulting nets, we find that about 5% nets violate timing constraints. Only these timing critical nets need to be ripped up and re-buffered by timing driven buffering techniques [1].

The main benefit from this new physical synthesis methodology is the huge gain in efficiency since slew buffering can be performed very efficiently. It is demonstrated in [1] (also in [3] for some similar results) that a slew driven buffering algorithm can run up to 88× faster than the timing driven buffering algorithm. For example, one can buffer 1000 industrial nets in only 6.2 seconds by the minimum cost slew buffering algorithm, while the minimum cost timing buffering algorithm needs 548.9 seconds.

This fast slew buffering algorithm proposed in [1] needs an important assumption, namely, the input slew to each buffer is assumed to be fixed at a conservative upper bound. With this input slew assumption, slew buffering can be efficiently performed under the dynamic programming framework. Certainly, improvement in buffer area is desired if this assumption is eliminated. As such, they [1] also propose a slew buffering algorithm without the fixed input slew assumption. The idea is to first discretize every possible input slew into slew bins and carry out the fixed input slew buffering algorithm with each bin. Since the solutions associated with a slew bin may be switched to other slew bins, numerous solutions can be generated. Experimental results in [1] show that although about 20% area saving can be obtained, the algorithm is not efficient: it is even slower than timing driven buffering in many cases. Thus, it is not attractive since a major reason for using slew buffering is its high efficiency. In order to make the approach practical, it is crucial to design a fast slew buffering algorithm for handling the non-fixed input slew case.

This work proposes a new fast slew buffering algorithm without input slew assumptions. In contrast to [1] which builds slew buffering solutions from scratch, we perform optimizations to buffering solutions obtained with the fixed input slew assumption. For this purpose, a heuristic is proposed to improve buffer usage under the slew constraint and it runs very fast. Together with the fact that slew buffering with fixed slew assumption can be efficiently computed, the whole approach runs very fast.

Our experimental results demonstrate the effectiveness and the efficiency of the new algorithm. Our algorithm runs up to 49× faster than the timing buffering algorithm with about 10% buffer area saving. Compared to the slew buffering algorithm proposed in [1], up to 37× speedup is obtained. Thus, our work makes the general slew buffering technique practical.

Note that there is another recent work in [4] which proposes a low-power buffering algorithm handling both timing constraint and slew constraint for timing critical nets. In contrast, the purpose of this paper is to address slew buffering on non-timing critical nets.

II. PRELIMINARIES

For completeness, we first introduce the slew problem as formulated in [1]. In the slew buffering problem, we are given a routing tree $T = (V, E)$. V consists of source vertex, sinks and internal vertices. Each sink has sink capacitance C_s. Each edge has lumped resistance R_e and lumped capacitance C_e. We are also given a buffer library B. Each type of buffer b has a cost W_b. At each internal vertex, some types of buffered can be inserted. A buffering solution is defined as a buffer assignment where buffers are inserted at some internal locations. The cost of a buffering solution γ is defined as

$$W(\gamma) = \sum_{b \in \gamma} W_b \quad [1].$$

We are to compute a minimum cost buffering solution such that the slew constraint is satisfied. The signal *slew* is the measure of rising or falling time of switching. As in [1], 10/90 slew is used which refers to the difference between the time signal waveform crosses the 90% point and the time signal waveform crosses the 10% point. Following [1], the slew model can be described by a generic example. Consider a path p from an upstream vertex u to a downstream vertex v. Assume that a buffer b is inserted at u and no buffer is inserted between u and v. Denote the output slew of b by $S_{b,out}(v)$ and the slew degradation along path p by $S_w(p)$. The slew $S(v)$ at v is computed as [5], [1]:

$$S(v) = \sqrt{S_{b,out}(v)^2 + S_w(p)^2}. \quad (1)$$

As the Elmore model for delay, the slew degradation along wire $S_w(p)$ can be computed by Bakoglu's metric [6] as

$$S_w(p) = \ln 9 \cdot D(p), \quad (2)$$

where $D(p)$ is the Elmore delay along p [1]. The output slew of a buffer, such as $S_{b,out}(v)$, depends on the input slew at this buffer and its load capacitance. As in [1], the dependence is described by a lookup table.

In [1], a fast algorithm is proposed to handle a simplified slew buffering formulation where the input slew to each buffer is assumed to be fixed at a conservative upper bound. This assumption allows us to process large number of nets very efficiently and slew constraint is satisfied. With the assumption, the output slew of buffer b at vertex v is then given by [1]

$$S_{b,out}(v) = R_b \cdot C(v) + K_b, \quad (3)$$

where $C(v)$ is the downstream capacitance at v, R_b and K_b are empirical fitting parameters. As in [1], we call R_b the slew resistance and K_b the intrinsic slew of buffer b.

The slew buffering problem is formulated in [1] as follows.

Slew Constrained Minimum Cost Buffer Insertion Problem [1]: Given a routing tree $T = (V, E)$, possible buffer positions, and a buffer library B, compute a buffering solution γ such that the total cost $W(\gamma)$ is minimized and the slew constraint α is satisfied.

III. OVERVIEW OF [1]'S MINIMUM COST SLEW BUFFERING ALGORITHM ASSUMING FIXED INPUT SLEW

The algorithm proposed in [1] works under the dynamic programming framework as timing buffering [7], [8]. For completeness, we include the algorithm in [1] as follows.

In the algorithm, a set of candidate solutions are propagated from the sinks toward the source. Each buffering solution γ is characterized by (C, W, S), where C denotes the downstream capacitance at the current node, W denotes the cost of the solution and S is the cumulative slew degradation along wire. S is S_w defined in Eqn. (2). The solution at a sink node has C as sink capacitance, $W = 0$ and $S = 0$. During solution propagation, we will perform "add wire", "add buffer" and "merge branch".

"Add Wire": to propagate a solution γ_v from a node v to its parent node u, a solution γ_u is generated at u as follows. $C(\gamma_u) = C(\gamma_v) + C_e$, $W(\gamma_u) = W(\gamma_v)$ and $S(\gamma_u) = S(\gamma_v) + \ln 9 \cdot D_e$, where $D_e = R_e(\frac{C_e}{2} + C(\gamma_v))$.

"Add Buffer": to insert a buffer b at u to γ_u, a new solution $\gamma_{u,buf}$ is generated as follows. $C(\gamma_{u,buf}) = C_{b_i}$, $W(\gamma_{u,buf}) = W(\gamma_u) + W_b$, and $S(\gamma_{u,buf}) = 0$.

"Merge Branch": at a branching node, two sets of solutions are merged. Denote by Γ_l the left-branch solution set and by Γ_r the right-branch solution set, respectively. For each solution $\gamma_l \in \Gamma_l$ and each solution $\gamma_r \in \Gamma_r$, the merged solution γ' is generated as follows. $C(\gamma') = C(\gamma_l) + C(\gamma_r)$, $W(\gamma') = W(\gamma_l) + W(\gamma_r)$ and $S(\gamma') = \max\{S(\gamma_l), S(\gamma_r)\}$.

The following pruning technique is performed to accelerate the approach in [1]. For any two solutions γ_1, γ_2 at the same node, γ_1 dominates γ_2 if $C(\gamma_1) \leq C(\gamma_2)$, $W(\gamma_1) \leq W(\gamma_2)$ and $S(\gamma_1) \leq S(\gamma_2)$. A solution is pruned if it is dominated by other solutions, or its cumulative slew degradation along wire $S(\gamma)$ is greater than the slew constraint α or the slew of any downstream buffer in γ is greater than α.

Refer to [1] for the further details and complexity analysis for the algorithm.

IV. A NEW FAST ALGORITHM WITHOUT FIXED INPUT SLEW ASSUMPTION

A. Problem of [1]'s Approach

In the above slew buffering algorithm, the buffer output slew is independent of its input slew since the input slew for each buffer is assumed to be fixed at a conservative upper bound. In this way, the feasible solutions returned by the above algorithm guarantee satisfying slew constraints. On the other hand, it is clear that if we eliminate the fixed-input slew assumption, there may be improvement in buffer area. This is demonstrated by experiments in [1] where 20% buffer savings can be obtained.

To handle slew buffering without fixed input slew assumption, [1] uses a much more complicated dynamic programming algorithm. Their idea is to first discretize every possible input

B. *The New Algorithm*

To handle non-fixed input slew case, instead of building buffering solutions from scratch, one might wonder whether we can start from a suboptimal yet easy-to-compute slew buffering solution and improve it through some heuristics. This motivates this work. Actually, such a solution can be easily obtained, which is just the one returned by the slew buffering algorithm with fixed input slew assumption as described in Section III. For convenience, we call such a solution an *initial solution*. Using slew lookup table, a top-down slew rate evaluation is first performed on the initial solution. It is possible that the input slews to some inserted buffers are much smaller than the slew constraints. Thus, we can try to replace some buffers with cheaper (smaller-cost) buffers. Buffer area saving is achieved if the slew constraint is still satisfied after the buffer replacement. This idea can be further generalized as follows. We do not need to restrict buffer replacement to be applied only at *initially buffered positions*, i.e., those positions with buffers inserted in the initial solution. More positions can be tried. Since our new algorithm is guided by the fixed input slew buffering algorithm, such positions should be those around the initially buffered positions.

Assume that a buffer is inserted at node v in the initial solution, then its immediate upstream and downstream $\pm P$ possible buffer positions will be investigated. We will first remove the buffer at v from the initial solution and tentatively insert a cheaper buffer into each of these $2P + 1$ positions. If the resulting solution is feasible (i.e., still satisfying the slew constraint), it will be recorded. The recorded solution with minimum cost will be selected as the buffer replacement solution for these $2P + 1$ positions. If there is no feasible buffer replacement, the initial solution will be restored. Since the upstream knowledge is available to us, the slew lookup table can be used for evaluating slew rate. Note that after processing any initially buffered position, only a single solution is maintained.

Our algorithm is now clear. After computing the initial solution, a local improvement heuristic is performed. For this, we compute a post-order traversal on the initially buffered positions. For each initially buffered position, perform the above buffer replacement process on those $2P + 1$ positions. After selecting the best buffer replacement solution for them, we proceed to the next initially buffered position along the post-order traversal. The process terminates after the last

buffered position is processed. Since for positions around each initially buffered position, only the solution with the minimum cost is selected (which certainly has no greater cost than the initial solution), the final solution is guaranteed to have no greater cost than the initial solution. In other words, this local improvement heuristic can never degrade the solution quality.

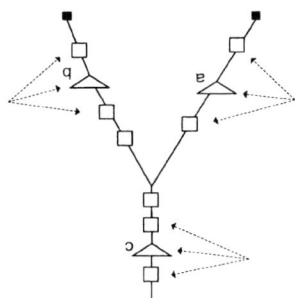

Fig. 1. An illustration of the local improvement heuristic. Square: candidate buffer position. Triangle: buffer.

Let us look at a simple example illustrating the above algorithm. Refer to Figure 1. In this tree, the initially buffered positions are a, b, c and they will be processed in this order. Suppose that P is set to 1. When a is processed, two adjacent possible buffer positions and a itself will be investigated. For each of the three positions, all buffers with cost smaller than the buffer cost in the initial solution (at a) will be tried. The minimum cost feasible solution will be finally selected after investigating all these three positions. b, c are then treated similarly. Note that in our algorithm, a restriction is further imposed: we require that all positions to be investigated must be along the same branch with the initially buffered position. For example, when $P = 2$, the position beyond the branching point is a position within $2P + 1$ around a. It will not be investigated as inserting buffer there will significantly change the initial solution (i.e., a buffer at a is removed while a buffer is inserted above the branching point). It should be avoided as our approach is guided by the initial solution. This strategy simplifies our task while still allows us to obtain high quality solutions as indicated by our experiments.

As regards to the time complexity of the local improvement heuristic, at most $(2P + 1)|B|$ slew validation (i.e., checking whether there is slew violation) are needed around each initially buffered position, where $|B|$ is the number of buffer types in the buffer library. Each slew validation can be performed in $O(n)$ time where n denotes the number of candidate buffer positions in the tree. This is due to that we need to perform slew evaluation at the node and all its downstream buffer positions. Therefore, there are slew violations, $(2P + 1)|B|$ slew validations need $O(n|P|B|)$ time. Denote by m ($m \le n$) the number of initially buffered positions, the runtime for the whole algorithm is bounded by $O(mnP|B|)$ time. As P is a constant, the runtime becomes $O(mn|B|)$. Note that in practice, the number of inserted buffers m is certainly much smaller than n. Thus, we can practically treat the above bound as $O(n|B|)$.

V. EXPERIMENTAL RESULTS

A. Experiment setup

To demonstrate the effectiveness and the efficiency of the new algorithm, we compare it with the approaches presented in [1]. For convenience, all algorithms in comparison are listed below together with their abbreviations.

- **SB**: slew buffering algorithm with fixed input slew in [1].
- **SB+NI**: slew buffering with non-fixed input slew in [1].
- **NEW**: new slew buffering with non-fixed input slew.
- **VGL+S**: slew constrained van Ginneken/Lillis' minimum cost timing buffering algorithm in [1].

For a fair comparison, our algorithm is tested on the same set of 1000 industrial nets as in [1].

B. Comparisons

For convenience, we first collect the results from [1] to Table I. Comparison of NEW with SB, SB+NI and VGL+S are summarized in Table II. Denote by "area saving," the percentage difference in area and by "speed up," the percentage difference in CPU time (seconds). The slew constraint is given in nanoseconds. In the experiment, P is set 1. Thus, three positions around each initially buffered position are investigated. As indicated in Section IV-B, these three positions must be along the same branch, and any position crossing the branching point will not be investigated since inserting a buffer there would significantly impact the initial solution. Since in our testcases, often there are no more than 3 candidate buffer positions along the same branch, $P = 1$ is a reasonable choice.

We make the following observations:

- Comparing NEW with SB, up to 10.8% buffer area saving is obtained by NEW with small amount of additional runtime. The computation overhead due to local improvement heuristic can be easily obtained by subtracting the runtime between NEW and SB. NEW and SB use the same number of buffers, which is the case since NEW only performs buffer replacement to the solutions obtained by SB and no buffers are further inserted or removed.

- Compared to SB+NI, NEW can be 37× faster. This is due to the fact that the local improvement heuristic can be efficiently performed and after it, only a single solution is selected while in SB+NI, the slew bin structure is used and solutions constantly switch bins so that a huge amount of solutions have to be propagated. As NEW runs much faster than SB+NI, the slight sacrifice in buffer usage becomes acceptable. NEW overcomes the hurdle of SB+NI for non-fixed slew buffering because it runs very fast. This matches our purpose for using slew buffering; slew buffering is for non-timing-critical nets, and thus runtime is the main issue.

- Comparing NEW with VGL+S, NEW can be 49× faster. Note that VGL+S runs slower with larger slew constraint. This is due to that VGL+S is a timing buffering and thus solution domination/pruning is based on slack but not on slew, and with larger slew constraint, fewer solutions can be eliminated from the solution set due to slew violation [1]. It is also clear that NEW often saves > 10% area over VGL+S.

TABLE I
RESULTS IN [1]. SLEW REFERS TO SLEW CONSTRAINT IN ns.

Slew	Fixed input Slew (SB)			Non-fixed Slew (SB+NI)			Timing Buf (VGL+S)		
	Area	#Buf	CPU	Area	#Buf	CPU	Area	#Buf	CPU
0.3	44980	7794	19.1	35148	7114	992.1	46551	6905	346.5
0.4	30960	6909	15.0	25010	5666	931.7	32133	7600	351.8
0.5	22960	5108	11.7	19797	4326	762.8	24235	6858	408.1
0.6	19380	4114	9.6	18528	3772	663.9	14838	4504	417.8
0.7	15531	3551	8.3	13995	3463	473.6	16445	4565	463.2
0.8	13340	3216	7.5	12129	3145	397.4	14218	4300	487.1
0.9	11578	2972	6.9	10667	2854	365.2	12243	3749	532.9
1.0	10316	2712	6.2	9626	2488	337.3	10807	3340	548.9

TABLE II
RESULTS OF NEW SLEW BUFFERING ALGORITHM WITH NON-FIXED INPUT SLEW. SLEW REFERS TO SLEW CONSTRAINT IN ns.

Slew	NEW			Ratios		
	Area	#Buf	CPU	AreaSaving SB	Speedup SB+NI	Speedup VGL+S
0.3	40128	7794	31.0	10.8%	32×	11×
0.4	28327	6909	25.2	8.5%	37×	14×
0.5	21761	5108	22.2	5.2%	35×	18×
0.6	17790	4114	17.7	3.2%	32×	24×
0.7	15017	3551	15.2	3.3%	30×	30×
0.8	12957	3216	15.1	2.9%	26×	32×
0.9	11582	2972	11.8	2.5%	31×	45×
1.0	10087	2712	11.2	2.2%	30×	49×

VI. CONCLUSION

Slew buffering can be used to efficiently buffer non-timing-critical nets and thus accelerate the whole physical design flow. However, the existing slew buffering algorithm proposed in [1] for handling non-fixed input slew runs slower than timing buffering which makes it less used in practice. This paper proposes a new fast slew buffering algorithm. Experimental results demonstrate that new algorithm can run up to 49× faster than [1] and up to 37× faster than the widely-used timing buffering algorithm with about 10% area saving.

REFERENCES

[1] S. Hu, C. Alpert, J. Hu, S. Karandikar, Z. Li, W. Shi, and C.-N. Sze, "Fast algorithms for slew constrained minimum cost buffering," *IEEE Transactions on Computer-Aided Design of Integrated Circuits and Systems*, accepted.

[2] P.J. Osler, "Placement driven synthesis case studies on two sets of two chips: hierarchical and flat," in *Proceedings of the ACM International Symposium on Physical Design*, pp. 190-197, 2004.

[3] S. Hu, C. Alpert, J. Hu, S. Karandikar, Z. Li, W. Shi, and C.-N. Sze, "Fast algorithms for slew constrained minimum cost buffering," in *Proceedings of ACM/IEEE Design Automation Conference*, pp. 308-313, 2006.

[4] Y. Peng and X. Liu, "Low-power repeater insertion with both delay and slew rate constraints," in *Proceedings of ACM/IEEE Design Automation Conference*, pp. 302-307, 2006.

[5] C.V. Kashyap and C.J. Alpert and F. Liu and A. Devgan, "Closed form expressions for extending step delay and slew metrics to ramp inputs," in *Proceedings of the International Symposium on Physical Design*, pp. 24-31, 2003.

[6] H.B. Bakoglu, *Circuits, Interconnects, and Packaging for VLSI*. Addison-Wesley Publishing Company, 1990.

[7] L.P.P. van Ginneken, "Buffer placement in distributed RC-tree networks for minimal Elmore delay," in *Proceedings of the IEEE International Symposium on Circuits and Systems*, pp. 865-868, 1990.

[8] J. Lillis and C.-K. Cheng and T.-T.Y. Lin, "Optimal wire sizing and buffer insertion for low power and a generalized delay model," *IEEE Journal of Solid State Circuits*, vol. 31, no. 3, pp. 437-447, 1996.

Closed form equations for inter-modulation distortion parameters in WCDMA receiver validated through measurements

Mohammed Saif Khan[1] and Naveen K. Yanduru[2]

[1]Texas Instruments, Bangalore, India. [2]Texas Instruments, Dallas, TX, USA.

Abstract — This work establishes the closed form equations that can be used to calculate the inter-modulation distortion noise in WCDMA receivers where the distortion mechanisms involve WCDMA modulated blockers. It is shown that the distortion noise depends not only on the power of the blockers but also on the number of channels in the modulated blocker. Simple equations have been provided which can be easily used by circuit designers in each case. The equations have been validated by a direct conversion receiver chip measurements as well as MATLAB simulations.

Index Terms — WCDMA, UMTS, 3G, RF receiver, distortion analysis, inter-modulation distortion.

I. INTRODUCTION

Most of the inter-modulation distortion mechanisms that affect the WCDMA receiver involve WCDMA modulated signals such as the TX (transmitter) leakage, WCDMA modulated blockers [1] (Fig. 1). The inter-modulation distortion that folds in-band depends on not only the power of the blockers but also on various factors such as the number of channels within the modulated blocker and the inter-modulation mechanism. One can easily observe that in these cases the standard equations that relate IP3 to IM3 and IP2 to IM2 cannot be used since they are more suited for CW tones and assume that the entire distortion falls in-band. Thus it is very important to establish this relation between inter-modulation distortion and circuit linearity parameters such as IP3 and IP2. This is also important because of the fact that RF and analog circuit designers typically design using SPICE based simulators which are not able to handle modulated blockers easily. Thus the required IP3 and IP2 for the receiver need to be equivalently "adjusted" so that circuit designers can simulate the linearity parameters using CW tones in SPICE.

This work focuses on establishing the equations for inter modulation distortion for various phenomenon encountered in WCDMA. The specific cases of TX leakage self mixing distortion, inter modulation between TX leakage and a modulated in-band blocker, inter modulation between a CW out of band blocker, cross modulation distortion between TX leakage and close-in blocker, and spectral re-growth due to adjacent channel will be discussed. All the equation

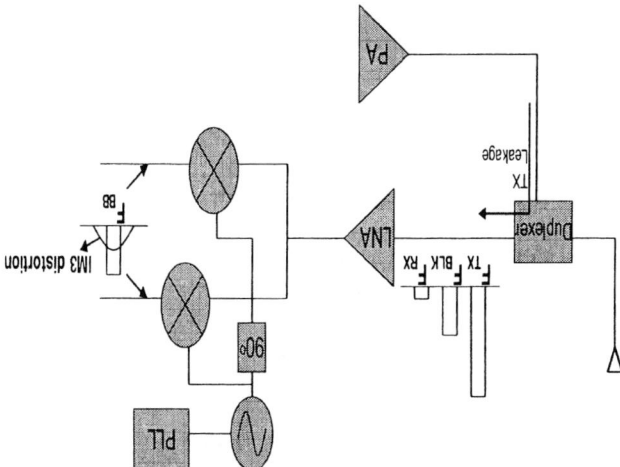

Fig. 1. Block diagram of a direct conversion WCDMA receiver showing inter-modulation between TX leakage and an in-band blocker.

II. TX SELF MIXING DISTORTION

In WCDMA system, there is a finite amount of TX leakage present at the RX input. Even the best duplexers do not provide more than 55dB of isolation and thus the TX leakage at the input of the RX (receiver) can be as high as -25dBm. The second order non-linearity of the direct conversion RF down converter can cause significant distortion. By using the appropriate signal model for the WCDMA modulated TX leakage, substituting it in the second order non-linearity model of the RF down converter, and by taking the fourier transform of the autocorrelation function of the output we get the power

provided are in dB scale. Detailed analysis on some of this phenomenon was reported previously by the same authors of this paper [2] [3] [4] [5]. In the case of TX leakage self mixing leakage distortion, the distortion power difference between the modulated analysis case and the CW standard IP2 based equation is more than 10dB which can make a big difference in the choice of circuit architecture decisions. In all closed form equations shown in subsequent sections, P_{TX} is the TX leakage power, IIP2 is the input referred IP2 and IIP3 is the input referred IP3.

spectral density of the distortion power [2] [6]. The distortion power within the down converted signal bandwidth can be calculated and represented in a format similar to the standard IP2 equation with an extra adjustment factor.

$$P_{IM2noise} = 2P_{TX} - IIP2 + Adj(N) \qquad (1)$$

Where, $Adj(N) = 10\log_{10}(3/8 - (7/24N))$

$P_{IM2noise}$ refers to the input referred inter-modulation distortion and N is the number of DPCH (Dedicated Physical Channels as defined in [1]) in the WCDMA TX leakage.

A direct conversion WCDMA receiver chip in 90nm CMOS was designed which includes an LNA and mixer. The CW IIP2 of the receiver using two distinct blockers separated by about 1MHz was first measured. Tests were then done with modulated WCDMA blocker with the number of DPCH channels (N) varied. Apart from the number of channels, another commonly used parameter to quantify the quality of the WCDMA modulated signal is the "Crest Factor". Crest factor seemed to provide better correlation between measured and theoretical results and is thus tabulated along with N in Table I. Crest factor is defined as the ratio of peak power (0.1% probability) to average power. Values of the adjustment factor from measurements and from theoretical calculations have been tabulated in Table I.

The adjustment factor in the calculation of self mixing caused by the TX leakage with one channel is -11.1dB measured and -10.8dB using equation (1).

TABLE I
Adj(N) VERSUS DPCH (IN BAND BLOCKING)

No of DPCH (N)	Crest factor (dB) (Min Crest factor setting in SMIQ03B signal generator [7])	Adj(N) (dB) (Measured)	Crest factor (dB) (Chip Measurements) (Theoretical)	Adj(N) (dB) (Theoretical)
1	4.8	-11.1	5.1	-10.8
2	4.8	-10.6	7.2	-6.4
3	6.0	-7.3	8	-5.56
4	6.4	-6.6	8.5	-5.2
5	7.0	-5.9	6.9	-5.0
9	7.0	-5.7	9.3	-4.86
16	8.5	-3.9	10.6	-4.48
32	9.1	-3.7	11.2	-4.36

III. THIRD ORDER INTER-MODULATION DISTORTION

All receivers suffer from various kinds of third order inter modulation distortion. FDD based systems such as WCDMA are more likely to suffer third order distortion due to the fact that the receiver already has a blocker in the form of TX leakage. One other blocker placed at an appropriate frequency can cause third order distortion. These cases are discussed in this section. The IP3 requirement of the RF front end from these phenomenons is dominated compared to the standard "inter modulation distortion" test case defined for WCDMA.

A. Inter modulation between TX Leakage and Modulated Blocker at (TX+RX)/2 MHz (In-Band Blocking Case)

As defined in 3GPP [1], the in band blocking test produces a scenario where a WCDMA modulated blocker could be located in the middle of the TX and RX frequencies (Fig. 1). The third order non linearity of the circuit can thus cause the inter modulation product to fall in band. Again, by using similar process for calculation of the distortion noise [3], we get the following equation:

$$P_{IM3} = 2P_B + P_{TX} - 2IIP3 + Adj(N) \qquad (2)$$

Where, $Adj(N) = 10\log_{10}(4/3 - 1/(2N))$

P_{IM3} refers to the input referred inter-modulation distortion, P_B is the power of the modulated blocker and N is the number of DPCH in the modulated blocker.

One important observation from equation (2) is that while the distortion noise depends on the number of DPCH in the modulated blocker it is not a function of number of DPCH in the WCDMA TX leakage.

TABLE II
Adj(N) VERSUS DPCH (IN BAND BLOCKING)

Number Of DPCH (N)	Blocker at (TX+RX)/2 Crest factor (Measured)(dB) (Average Crest factor setting in SMIQ03B)	Adj(N) (Measured)(dB)	Blocker at Crest factor (Theoretical result)(dB)	Adj(N) (Theoretical result)(dB)
1	4.8	-0.87	5.1	-0.79
2	4.8	-0.39	7.2	0.35
4	6.6	0.69	8.5	0.82
6	8.2	1.44	9.3	0.97
16	9.3GPP OCNS[1]	1.66	10.3	1.15
12	11.2	1.97	10.1	1.11

B. Inter-modulation between TX Leakage and CW Blocker at (2*TX - RX) MHz (Out of Band Blocking Case)

As defined in 3GPP [1], the out of band blocking test creates a scenario where a CW blocker is located at a frequency of $(2*TX - RX)$. The inter modulation distortion noise [3] is given by

$$P_{IM3} = 2 \cdot P_{TX} + P_B - 2IIP3 + Adj(N) \quad (3)$$

Where, $Adj(N) = 10log_{10}(3/2 - 7/(12N))$

P_{IM3} refers to the input referred inter-modulation distortion, P_B is the power of the CW blocker, and N is the number of DPCH in TX leakage.

TABLE III
ADJ(N) VERSUS DPCH (OUT OF BAND BLOCKING)

Number of DPCH in the TX leakage	TX Crest factor (Chip generator) Setting in SMIQ03B (dB) (Min Crest Factor)	Adj(N) (dB) (Measurement)	TX Crest Factor (Measurement) (dB)	Adj(N) (dB) (Theoretical Result)
1	4.8	-0.38	5.1	-0.38
2	4.8	0.19	7.2	0.82
3	6.0	0.95	8.0	1.16
4	6.5	1.12	8.5	1.32
9	7.0	1.24	9.3	1.47
8	7.2	1.43	9.87	1.55
16	8.0	1.63	10.3	1.65

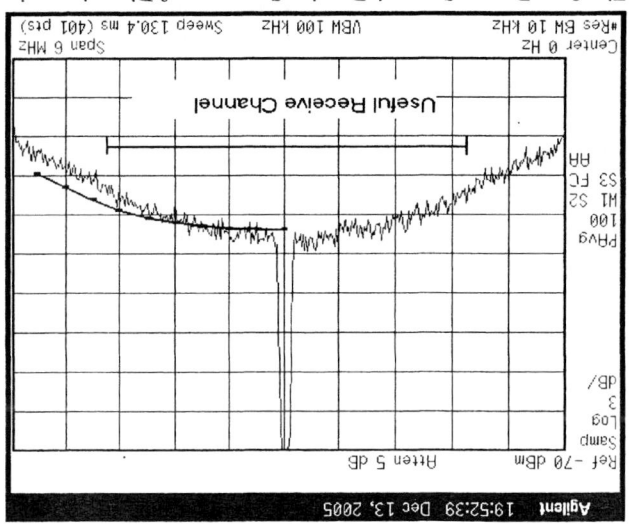

Fig. 2. Power Spectral Density Spectrum of Distortion noise ("In band blocking case") with 2 DPCH channels in the blocker at (TX+RX)/2 Hz. Dark bold line represents the spectrum as per the theoretical result given by equation (2).

IV. CROSS MODULATION DISTORTION

A. CW blocker close to signal band

As defined in 3GPP [1], narrow band blockers could be located very close to the RX carrier. In these cases cross modulation distortion will happen in the presence of the WCDMA modulated TX leakage. The cross modulation distortion noise can be calculated [4] as

$$P_{xmod\_cw} = 2P_{TX} + P_B - 2IIP3 + Adj(N) \quad (4)$$

When $1/2 < |w_b - w_{rx}|/2\,W < 1$
$Adj(N) = 10log_{10}(4[(1/2 - 1/(3N)) - u(1 - 1/N) + u^2(1/2 - 1/N) + u^3/(3N)])$
Where, $u = (|w_b - w_{rx}|/2\,W - 1/2)$
and
When $1 < |w_b - w_{rx}|/2\,W > 3/2$
$Adj(N) = 10log_{10}([2y^2 + 4y^3/(3N)])$
Where, $y = (|w_b - w_{rx}|/2\,W - 3/2)$,

When $|w_b - w_{rx}|/2\,W > 1/2$, the blocker is inside the useful channel, and therefore evaluation of cross-modulation distortion power is not valid. When $|w_b - w_{rx}|/2\,W > 3/2$, the entire distortion power falls out of the RX signal band and hence P_{xmod_cw} is zero.

In the above equations, P_{xmod_cw} refers to the input referred cross modulation distortion noise, P_B refers to the blocker power, N is the number of DPCH in WCDMA TX leakage, w_{rx} is the angular frequency of the RX carrier and W is the angular frequency of the blocker, w_b is the leakage signal bandwidth of 3.84MHz.

Using the above formula the Adj(N) for cross modulation distortion in equation (4) for the 2.7MHz narrow band blocking case of WCDMA is -2.27dB for a single DPCH in TX leakage. The measured result in this case for Adj(N) is -3dB.

B. Modulated blocker close to signal band

Cross modulation distortion can also happen due to a 5MHz offset adjacent channel WCDMA blocker. The distortion noise calculated [4] in this case is given by

$$P_{xmod} = 2P_{TX} + P_B - 2IIP3 + Adj(N) \quad (5)$$

Where, $Adj(N) = 10log_{10}(u^3(2 - u/N)/3), u=0.698$

P_{xmod} refers to the input referred cross modulation distortion noise, P_{TX} is the TX leakage power, P_B is the input WCDMA modulated blocker power, IIP3 is the input referred IP3 and N is the number of DPCH in TX leakage.

An important observation from equation (5) (for a single DPCH TX Leakage signal) is that the distortion power is independent of the number of DPCH in the modulated blocker and only depends on the overall power of the modulated blocker. Calculated Adj(N) from equation (5) for the 5MHz adjacent channel case is -8.32dB against the measured result of -9dB.

V. ADJACENT CHANNEL SPECTRAL RE-GROWTH

Distortion noise from spectral re-growth of the adjacent channel (5MHz) has also been analyzed [5].

$$P_{srg} = 3P_{ac} - 2.IIP3 + K_{dB} \qquad (6)$$

Where P_{srg} is the input referred spectral re-growth distortion power, P_{ac} is the blocker power, IIP3 is the input referred IP3 and K_{dB} is a factor which is dependent on the number of DPCH of the adjacent channel. Closed form equation showing the value of K_{dB} is not shown but the fourier transform of the auto-correlation function is coded in MATLAB. The measured results for K_{dB} and MATLAB simulation results are shown in Tables IV and V respectively.

TABLE IV
KDB VERSUS CREST FACTOR (LAB MEASUREMENT)

No OF DPCH	Crest factor	K_{dB}
1	4.8	-13.75
3	6.0	-10.3
4	6.6	-9.47
5	7.9	-7.86
6	8.2	-7.19
8	9.1	-4.91
9	10.0	-4.42
10	10.7	-3.75
14	11.4	-3.54
3GPP OCNS[4]	9.2	-6.02

TABLE V
KDB VERSUS CREST FACTOR (MATLAB MEASUREMENT)

No OF DPCH	Crest factor	K_{dB}
1	5.15	-12.13
2	7.2	-7.77
3	8.0	-6.41
4	8.5	-5.63
5	8.93	-5.15
6	9.28	-5.02
9	9.87	-4.53
12	10.06	-4.27
15	10.2	-4.19

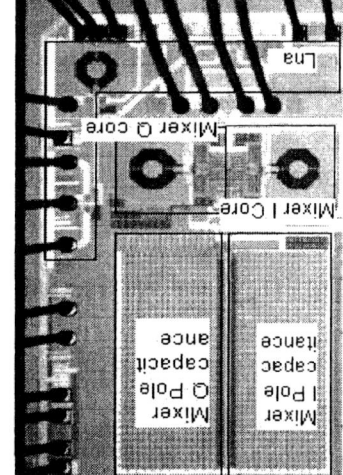

Fig. 3. Zoomed in Photograph of the Receiver die.

VI. CONCLUSION

The significance of PAR in the intermodulation distortion caused by AM blockers is highlighted. Mathematical equations for various intermodulation scenarios for WCMA receiver have been provided and are shown to have good agreement with time domain simulations and chip measurements. The analysis and results also shows the importance of such analysis compared to first order approximation done using CW based formulas. The results can be easily integrated into receiver system models for developing IC specification, performing system validation and budgeting. Though the results have been shown for WCDMA standard, the concept can easily be extended to other standards.

REFERENCES

[1] 3GPP TS 25.101, UE (User Equipment) radio transmission and reception. www.3gpp.org

[2] Mohammed Saif Khan and Naveen Yanduru, "Analysis and Measurement Of Self Mixing Of Transmitter Leakage in WCDMA Receivers," RAWCON 2007.

[3] Mohammed Saif Khan and Naveen Yanduru, "Analysis of Signal Distortion due to Third Order Non-Linearity in WCDMA Receivers," IEEE ISCAS 2006.

[4] Mohammed Saif Khan and Naveen Yanduru, "Analysis and Measurement of Cross Modulation Distortion in WCDMA Receivers," to be published in IEEE APCCAS 2006.

[5] Mohammed Saif Khan and Naveen Yanduru, "Signal Distortion due to Spectral Re-Growth of Adjacent Channel Interferers in WCDMA Receivers," RAWCON 2007.

[6] W. Y. Ali Ahmad, "Effective IM2 estimation for two-tone and WCDMA modulated blockers in zero-IF receivers", RF Design, pp. 32-40, April 2004.

[7] Operating Manual, Rohde-Schwarz SMIQ03B Vector Signal Generator, Volume 1, Section 2.14.3.4.

A CMOS Wideband LNA Using Multiple Phase Matched Frequency Staggered Resonators

Diptendu Ghosh and Ranjit Gharpurey

Department of Electrical and Computer Engineering, University of Texas at Austin, Austin, TX 78712, USA

e-mail: {diptendu, ranjitg}@mail.cerc.utexas.edu

Abstract – A new architecture of wideband LNA has been introduced. High gain is obtained through use of narrowband tuned stages and broadband performance is simultaneously achieved by using multiple of them, with their resonant frequencies staggered, in parallel. Multipath phase matching is shown to be effective in optimizing passband gain and NF with respect to broadband noise performance over conventional cascaded narrowband matching network based designs and considerable power reduction at similar gain in comparison to typical distributed amplifiers. Simulations in 0.13um CMOS process show 22.3 dB peak gain, 38% fractional bandwidth about 26.5-GHz, 2.8-3.8 dB NF over entire band and mid-band output P1dB of 2.6 dBm. It consumes 31 mA from 1.5 V supply.

Index Terms – high frequency front end, low noise amplifier, distributed, phase matching, frequency staggering, parallel resonators, power combiner

I. INTRODUCTION

WITH the evolution of new ultra-wideband standards for high data rate communication lot of research is being done towards development of wideband LNAs. Narrowband amplification at frequencies close to device F_T is possible using inductors for resonating parasitic device capacitances and simultaneous matching for noise and power using inductive degeneration techniques [1, 2]. Unfortunately for wideband case getting a good noise and power match over large band of frequencies makes the problem quite challenging. Added to this is the scarcity of power gain available from active devices at multi-GHz frequencies and passive loss mechanism which worsens at high frequencies.

This paper describes a wideband LNA operating above 10-GHz in short-channel CMOS technology. Section II reviews the existing approaches for wideband LNA design and shows their drawbacks. In Section III our proposed architecture is detailed. Section IV explains the power advantage of our architecture over other distributed approaches [3, 4]. Performance optimization using multipath phase matching is elaborated in Section V. Simulation results are presented in Section VI.

II. EXISTING APPROACHES FOR BROADBAND LNA DESIGN

Most of the existing approaches towards designing broadband LNA have gone towards the realization of amplifiers embedded within broadband matching networks (BWM) [5]. The frequency dependent V_{gs} applied to critical noise contributing transistor in these topologies results in a degradation of the noise figure by significant amount e.g. 4-5 dB at the band edges. Besides, trying to extend these

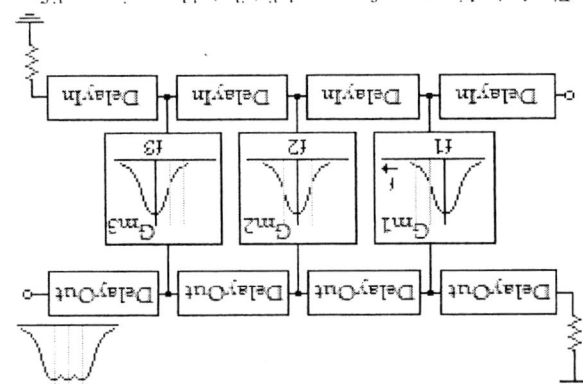

Fig. 1. Architecture of proposed distributed low noise amplifier

techniques to frequencies beyond 10-GHz, e.g. the 22-29-GHz band, leads to requirements of high-Q inductances in the order of nH which is difficult to realize due to poor self-resonant frequencies(f_{sr}) of large inductors.

Other conventional approaches deal with using cascaded frequency staggered gain stages and narrowband input, inter-stage and output matching networks(NBMN) to achieve high gain [6]. Here also, the structures suffer from sharp rise in NF beyond the resonant frequency of the input matching network. Besides, they do not have good broadband s_{11}.

Broadband amplification from dc to near 100-GHz in ultra-deep sub-micron technologies have been reported recently [7] - exploiting the advantages of distributed amplification (DA) [8]. Although suffering from higher power dissipation and low gain their noise performance is better over wide bandwidth. However, work towards getting wideband amplification from DA with non-zero lower cutoff frequency(BPDA) has been limited to incremental changes of the input-output matching network. Recently [3, 4] reported such implementations where the matching network component values were chosen to realize approximate higher order filter transfer functions. These approaches have limited theoretical rigor also leading to transfer function modification and worsening of frequency response. Moreover, such techniques which use passive input and output frequency shaping networks and broadband g_m-cells are, as explained in Section IV, not optimal in terms of power performance for similar gains.

III. PROPOSED ARCHITECTURE

The proposed architecture is shown conceptually using the block diagram in Fig. 1. It comprises of a number of parallel narrowband transconductors realized by using stagger-tuned components. The architecture is based on "distributed" amplification in the sense that it exploits signal propagation along multiple parallel paths working in synchrony to get performance advantages [9]. The main difference from existing DA topologies comes from using narrowband stages for high gains at discrete frequencies within the passband and simultaneously getting wide bandwidth from superposition of frequency staggered narrow band currents coming from the different phase matched paths. The motivation for staggering is to widen the flat gain region without compromising the out-of-band rolloff or passband gain. Broadband matching is obtained through absorption of device capacitances into input and output TLines.

Schematic for the implementation is shown in Fig.2. The transconductors are 2-staged with the first stage device input capacitance ($\sim C_{gs,M1}+C_{gd,M1}$.MillerFactor) absorbed in the input matching network. The output capacitance of the first stage ($\sim C_{db,M1C}+C_{dg,M1C}$) along with the input capacitance of the second stage ($\sim C_{gs,M2}+C_{gd,M2}$.MillerFactor) is tuned to resonate with an inductive load (L_{load}) at a frequency within the passband gain. This ideally produces a frequency shaped passband gain g_m's as the second voltage being fed to infinitely broadband stage output capacitance ($\sim C_{db,M2C}+C_{dg,M2C}$) is made a part of the output TLine. The second stages of the g_m-cell and output line thus constitute an active power combining network. The total phase shift along each path having 3 segments – the input TLine, the intermediate narrowband stage and output TLine – is approximately matched across paths. This, as explained in Section V, is important for having good passband flatness and for improving inband NF.

DC blocking capacitor (C_I) is required to have independent biasing of the 2 stages. Cascoding (M1C, M2C) is required for reverse isolation and also to prevent degradation of the frequency response due to coupling of the voltage from high gain nodes to the input lines. Inter-stage matching inductors (L_I) [10] are used for boosting the power gain of the cascode stage. An inductor L_{SH} shunts the load-termination resistor (R_{LT}) to keep the output bias point at supply and this helps increasing linearity. The value of L_{SH} is lower limited by the inband s_{22} degradation it produces. m-derived sections are used for improved bandwidth performance of the input and output lumped element TLines. An L-match (L_M, C_M) is used for matching the output transmission line with $Z_0 = 100\text{-}\Omega$ to

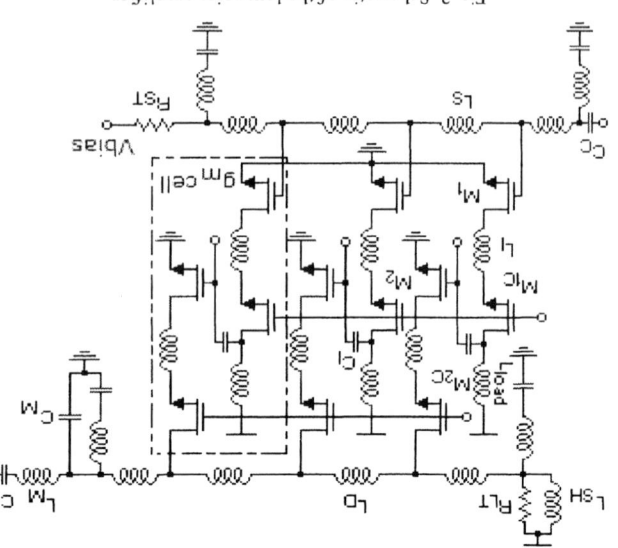

Fig. 2. Schematic of the low noise amplifier

the 50-Ω load. Placing an L-match at the output improves input gain within the bandwidth of interest but do not compromise noise greatly due to its placement after the high gain stages.

IV. POWER ADVANTAGE OVER DA BASED APPROACHES

To demonstrate the power advantage of this topology we now make comparison with conventional BPDA architectures [3, 4] which shape frequency response using passive filters at input and output and use transistors for synthesizing the broadband g_m(s). In the comparison, which follows, it is to be noted that bandwidth realization is through different means in the two cases – and we are assuming that the requirement is met in both methods. For consistency the number of staggered paths (n_{sl}) is taken as identical to the number of transistors (n) in the conventional DA. If $n_{sl} > n$ then we will not meet bandwidth specification in the new architecture; but if it is less then we will get more power advantage. $g_{m,conv}$, g_{m1} and g_{m2} represent the transconductance of conventional and 2 stages of our gm-cell respectively. R is the load of final stage and R_L is our first stage load.

Conventional BPDA mid-passband gain $\leq$

$$\frac{n \cdot g_{m,conv} R}{2} \quad (1)$$

R is typically fixed to 50-Ω. n_{max} is limited to 4 for CMOS implementations due to passive loss mechanisms.

Considering a 12 dB typical gain $g_{m,conv} R \big|_{n=4} = 2 \quad (2)$

Staggered BPDA mid-passband gain ignoring the L-match and gain improvement from superposition across paths =

$$\frac{g_{m,eq} R}{2} = \frac{g_{m1} R_L g_{m2}}{2} R = (g_{m1} R_L)\frac{g_{m2} R}{2} \quad (3)$$

Even for a conservative inductor Q ($\sim$6) R_L is higher than 200 $\Rightarrow g_{m1} R_L = 8$ if we burn same amount of current in M_1 as in conventional DA transistor. This follows from (2). From the above we can see that to get the same gain with conventional architecture having 4 transistors – $g_{m,conv}$ has to be boosted by 2x with respect to $g_{m,new} = g_{m1} = g_{m2}$. $g_{m,conv} = 2g_{m,new}$ is not achievable by burning more current if we are in the region operating at peak f_T of the process. However, if we are in the region significantly below peak f_T then $g_m \propto \sqrt{I_D}$ i.e. Power ∞ $I_D = f(g_m) \propto g_m^2$. Thus we can infer

$$\frac{Power_{conv}}{Power_{new}} = \frac{n\, f(g_{m,conv})}{2 n_{sl}\, f(g_{m,new})} = \frac{f(g_{m,conv})}{2 f(g_{m,new})} = \frac{(2g_m)^2}{2(g_m)^2} = 2 \quad (4)$$

Under aforesaid assumptions, conventional topology requires at least 2x power to achieve identical gain and bandwidth. We can, therefore, conclude in general that narrowband staggering approach is advantageous since "g_m" cells trade bandwidth with gain and thus gives higher equivalent g_m over narrow bandwidth at lower current. Bandwidth can be regained using staggering even at lower total power budget.

V. PHASE MATCHING AMONG PATHS IN N-PATH BPDA

In the following derivations, we ignore effects of L-match, cascading and Tline loss to avoid complicacy. Forward Gain (FGain) and Reverse Gain (RGain) refer to transfer function from source and source termination resistor (R_{ST}) respectively to load (R_L). Also, we index the parameters of N paths of the

BPDA with k=1(leftmost in Fig. 1) to N. $\{L_d, L_g, L_k\}$ and $\{C_d, C_g, C_k\}$ are the inductances and capacitances of the drain, gate transmission lines and k^{th} path tank respectively. Therefore,

$$\text{Forward Gain} = \frac{1}{2}R_L \sum_{k=1}^{N} g_{m1}g_{m2}\frac{\dfrac{C_k}{s}}{s^2 + \dfrac{\omega_k}{Q_k}s + \omega_k^2}\,e^{-\beta_g(k)}e^{-\beta_d(N-k+1)} \quad (5)$$

$$\text{Reverse Gain} = \frac{1}{2}R_L \sum_{k=1}^{N} g_{m1}g_{m2}\frac{\dfrac{C_k}{s}}{s^2 + \dfrac{\omega_k}{Q_k}s + \omega_k^2}\,e^{-\beta_g(N-k+1)}e^{-\beta_d(N-k+1)} \quad (6)$$

$$\text{where,}\qquad \beta_g = s\sqrt{L_g C_g} = s/\omega_g \quad (7)$$

$$\omega_k = \frac{1}{\sqrt{L_k C_k}} = k^{th}\text{ Path resonance frequency} \quad (8)$$

$$Q_k = \omega_k R_k C_k \quad (9)$$

We assume $R_k C_k$ to be same for all resonators i.e. $Q_k = \omega_k RC$. This is realistic if we are having same device size (M_2) for all the parallel paths and maintain same peak gain in each of them.

The transconductor in the kth path has the transfer function

$$G_{m,k} = \frac{\dfrac{C_k}{g_{m1}g_{m2}}\,s}{s^2 + \dfrac{\omega_k}{Q_k}s + \omega_k^2} \quad (10)$$

Thus, $\angle G_{m,k}(j\omega) = 90^\circ - \tan^{-1}\dfrac{\dfrac{\omega_k}{Q_k}\omega}{\omega_k^2 - \omega^2}$ (11)

At frequencies close to the resonant frequencies ω_k

$$\angle G_{m,k}(j\omega) = 2\frac{Q_k}{\omega_k}(\omega_k - \omega) = 2\frac{Q_k}{\omega_k}\Delta\omega = 2RC\Delta\omega \quad (12)$$

The phase shift along k^{th} path for FGain, from (5) and (12), is

$$\theta_{total,k} \approx k\omega\sqrt{L_g C_g} - (N-k+1)\omega\sqrt{L_d C_d} - 2R_k C_k(\omega - \omega_k)$$
$$= -k\frac{\omega}{\omega_g} - (N-k+1)\frac{\omega}{\omega_d} - 2RC(\omega - \omega_k) \quad (13)$$

Using (6) and (12) RGain phase shift is

$$\theta_{total,k\text{-term}} = -(N-k+1)\frac{\omega}{\omega_g} - (N-k+1)\frac{\omega}{\omega_d} - 2RC(\omega - \omega_k) \quad (14)$$

For phase, we make the approximation of matching only at a number of critical frequencies which are the geometric mean of $(N-1)$ adjacent resonant frequency pairs. Also, we make the assumption that at those frequencies only the paths whose tank resonant frequencies are closest superpose and contribute

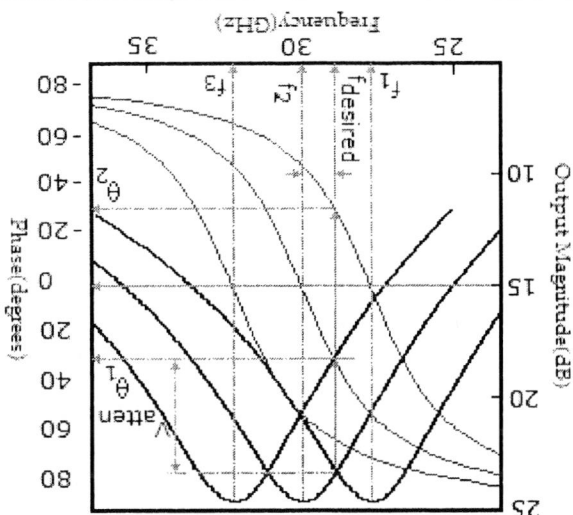

Fig. 3. Gain and phase response for a number of staggered resonators

most to signal amplitude. Other paths provide enough relative attenuation such that that phase imbalance caused by any of them is minimal on resultant transfer function. The amount of staggering, given by ω_{k+1}/ω_k, thus should be such that under given Q the "incrementally" linear phase relationship (12) is valid for 2 adjacent paths but high enough to make amplitude assumption for signals arriving through non-adjacent paths justified. Fig. 3. shows critical parameters involved in above derivation using 3 staggered resonators, each with Q of 10. In context of this figure the region between $\{f_1, f_{desired}\}$ and $\{f_{desired}, f_3\}$ has to be linearly varying with phase and V_{atten} should be large. Violating these simplifying assumptions midly, however, will not alter the general conclusions drawn.
Thus from (13), for phase matching among adjacent paths

$$\Delta\theta_{total,k\to k+1} = \theta_{total,k} - \theta_{total,k+1} = \left.\left(\frac{\omega}{\omega_g} - \frac{\omega}{\omega_d}\right) + 2RC(\omega_k - \omega_{k+1})\right|_{\omega=\omega_{desired}} = 0$$

$$\text{where } \omega_{desired} = \sqrt{\omega_k\,\omega_{k+1}} \quad (15)$$

$$2RC\left(\frac{\omega_k}{\omega_{k+1}} - 1\right) = \left(\frac{1}{\omega_g} - \frac{1}{\omega_d}\right) = 2RC\left(\frac{\omega_k}{\omega_{k+1}} - 1\right)\left(\frac{2}{3} + \frac{\varepsilon}{2}\right) \approx 2RC\,\varepsilon \approx 2\frac{Q_{desired}}{\omega_{desired}}\varepsilon \quad (16)$$

In cases where it is not possible to satisfy (16) exactly, we can still improve performance by maximizing forward to reverse superposition gain ratio

$$\frac{\cos\!\left(\dfrac{\Delta\theta_{total,k\to k+1}}{2} - \omega_{desired}\!\left(RC\varepsilon + \dfrac{1}{2\omega_g} + \dfrac{1}{2\omega_d}\right)\right)}{\cos\!\left(\omega_{desired}\!\left(RC\varepsilon + \dfrac{1}{2\omega_g} + \dfrac{1}{2\omega_d}\right)\right)} \quad (17)$$

Simulations reveal that due to low Q requirements imposed by (16), optimizing (17) leads to better overall performance.

Improved phase matching is required for reducing passband amplitude and group delay ripple. It also helps increasing the forward to reverse gain ratio which thereby improves the noise performance. To demonstrate the validity of our theory Fig. 4 shows the forward and reverse gain of the amplifier for the 2 possible monotonic ordering of the resonant frequencies of the narrowband transconductors. Fig. 5 shows the NF in the two cases. It is clearly seen from the figures that the resonant frequency order of $f_{reso1}<f_{reso2}<f_{reso3}$ (i.e. positive ε in (16)) is best both in terms of the FGain/RGain ratio and noise due to better phase matching. It is worth noting that typically gate capacitances are larger than drain ones making $\omega_d > \omega_g$ and thereby implying that a positive ε gives better match at constant resonator Q. The primary cause of NF degradation is

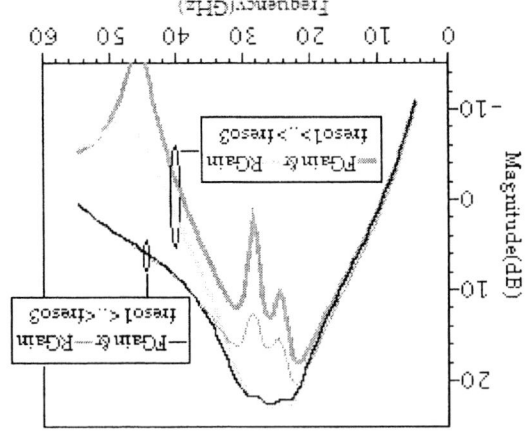

Fig. 4. Forward and Reverse Gain for different ordering of resonators

the source termination resistor due to high RGain/FGain in resonance frequency ordering, $f_{reso1}>f_{reso2}>f_{reso3}$. The passband response is also noticeably worse in latter case.

VI. SIMULATION RESULTS

Simulation results for the proposed LNA in 0.13um CMOS technology are shown in Fig. 6 and 7. The performance is summarized in the numbers below.

Supply = 1.5V, Current = 31mA
Gain > 21dB, Inband ripple < 0.5 dB
Other S-parameters: s_{11} <-14dB, s_{22}<-8dB s_{12} <-60dB
NF < 3.8dB, NF_{min} = 2.8dB
OP1dB at 27-GHz = 2.7dBm
IIP3(tones at 26 and 26.25-GHz) = -10.6dBm
3-dB Bandwidth = 38%

It can be clearly seen that our architecture gives much better broadband noise performance when compared to [5]. Results in [6] also show widely changing NF. The gain obtained here is much higher than other distributed approaches. It is noted that to get similar gains power requirement would have gone up significantly in the DAS [7].

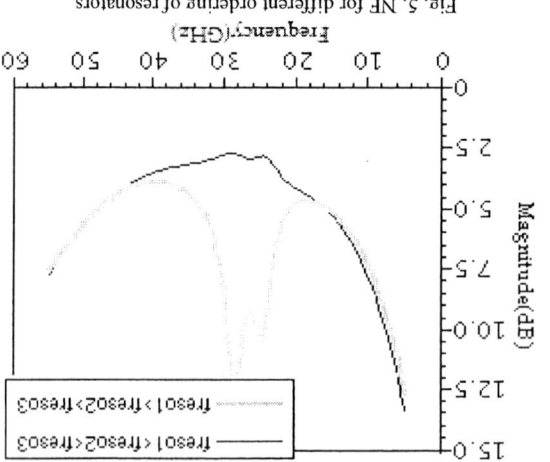

Fig. 5. NF for different ordering of resonators

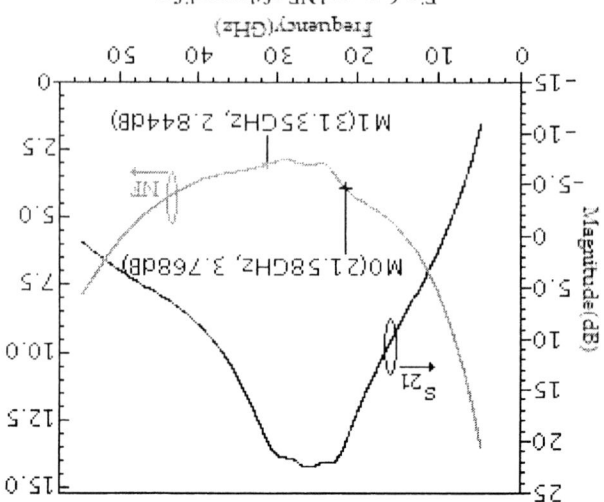

M1(3.35GHz, 2.844dB)
M0(21.58GHz, 3.768dB)

Fig. 6. s_{21} and NF of the amplifier

VII. CONCLUSION

A new architecture of broadband LNA in scaled CMOS process working at frequencies in excess of 10-GHz has been presented. Multiple frequency staggered parallel resonators are used for getting high gain and wide bandwidth at lower power than conventional distributed amplifiers. Multipath phase matching is used to improve performance. The topology gives truly broadband noise performance even at high frequencies. Simulations in 0.13um CMOS process show a 22.3 dB peak gain, 38 % fractional bandwidth about 26.5-GHz, 2.8 to 3.8 dB NF over entire band, mid-band output P1dB of 2.6 dBm. It consumes 31mA from 1.5V.

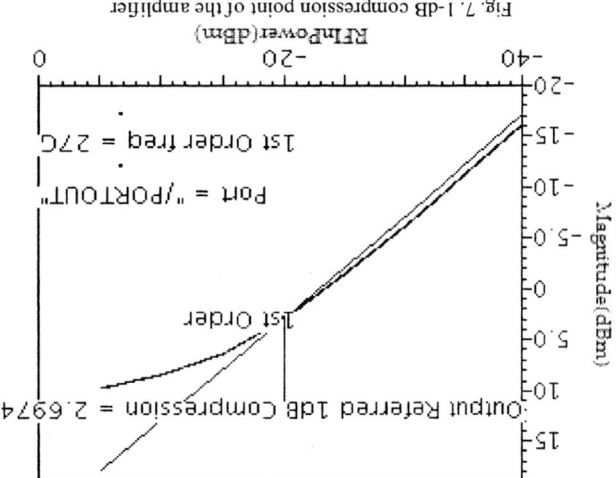

Fig. 7. 1-dB compression point of the amplifier

REFERENCES

[1] T. H. Lee, *The Design of CMOS Radio-Frequency Integrated Circuits*, 1st ed. Cambridge University Press, 1998.

[2] B. Razavi, "A 60-GHz CMOS receiver front-end", *IEEE J. Solid-State Circuits*, vol. 41, no. 1, pp. 17 - 22, Jan. 2006.

[3] K. K. Moez and M. I. Elmasry, "Design of broadband bandpass CMOS amplifiers based on modified distributed amplification technique", in *Proc. IEEE Int. MWSCAS*, 2005, pp. 794-797.

[4] N.P. Mehta and P.N. Shastry, "Design guidelines for a novel bandpass distributed amplifier", in *Proc. European Microwave Conference*, 2005, vol. 1.

[5] A. Bevilacqua and A.M. Niknejad, "An ultra-wideband CMOS LNA for 3.1 to 10.6 GHz wireless receivers", in *IEEE ISSCC Dig. Tech. Papers*, 2004, pp. 382 - 383.

[6] C.H. Doan, S. Emami, A.M. Niknejad and R.W. Brodersen, "Millimeter-wave CMOS design", *IEEE J. Solid-State Circuits*, vol. 40, no.1, pp. 144-155, Jan. 2005.

[7] M. Tsai, H. Wang, J. Kuan and C. Chang, "A 70GHz cascaded multi-stage distributed amplifier in 90nm CMOS technology", in *IEEE ISSCC Dig. Tech. Papers*, 2005, pp. 402 - 403.

[8] C.S. Aitchison, "The Intrinsic Noise Figure of the MESFET Distributed Amplifier", in *IEEE Trans. Microw. Theory and Tech.*, vol. 33, no. 6, pp. 460 - 466, June 1985.

[9] A. Hajimiri, "Distributed integrated circuits: an alternative approach to high-frequency design", *IEEE Communications Magazine*, vol. 40, no. 2, pp. 168 - 173, Feb. 2002.

[10] B. Analui and A. Hajimiri, "Bandwidth enhancement for transimpedance amplifiers", *IEEE J. Solid-State Circuits*, vol. 39, no. 8, pp. 1263 - 1270, Aug. 2004.

A Technique to Extend Tuning Range of High Frequency Quadrature VCO

Diptendu Ghosh and Ranjit Gharpurey

Department of Electrical and Computer Engineering, University of Texas at Austin, Austin, TX 78712, USA

e-mail: {diptendu, ranjitg}@mail.cerc.utexas.edu

Abstract – A new approach to extend the tuning range of high frequency quadrature VCO has been proposed. The method exploits the existence of two possible modes of operation, with distinct frequencies, in LC QVCOs based on a pair of coupled cores. Mode switching is achieved by choosing between two types of reactive degeneration in coupling transistors. Used along with varactor-control-based continuous tuning, in an implementation in 0.13um CMOS process, the technique extends tuning range by more than 85% over varactor-only based approach at similar oscillation frequency, phase noise, power and quadrature accuracy. This method also has limited noise degradation across the entire tuning range and requires fixed amount of power for all frequencies. Consuming a total current of 11mA from 1.2V supply the QVCO exhibits simulated phase noise of -104.7 to -106.6dBc/Hz at 1MHz offset from center frequency. The tuning range is from 17.78 to 20.64-GHz.

Index Terms – high frequency VCO, discrete tuning, coupled oscillator, reactive degeneration, dual mode

I. INTRODUCTION

HIGH frequency voltage controlled oscillator is one of the key building blocks of contemporary communication systems. It is used in wireless transceivers to perform open or closed loop frequency synthesis, in optical communication systems for high speed clock and data recovery(CDR). Quadrature VCOs are particularly useful in direct conversion as well as image reject receivers and half-rate CDRs [1, 2].

Rapid technological advancements have rendered low cost CMOS technologies amenable to the implementation of QVCOs at millimeter-wave frequencies. However, these high frequency designs in CMOS typically suffer from worser performance and very stringent tradeoffs between the different critical specifications. For instance, in LC based oscillators, large device capacitance and low self-resonance frequencies(f_{sr}) of high-Q inductors impose significant limitations to tuning capability for operation at frequencies close to device f_T.

Wide tuning range(TR) is essential for counteracting process variations which is acute in deep-submicron technologies. Moreover, wideband applications require an agile vco to sweep across the frequency range defined by the standard and consequently a high TR. Thus several approaches have been investigated in increasing tuning range — most trading some other performance metric to achieve this end. In this paper we investigate a new technique which imposes no power or phase noise(PN) penalty to get TR enhancement. The paper is organized as follows – Section II briefly reviews existing techniques to widen TR and their limitations. In Section III we describe our proposed technique and analyze the mode switching behavior in Section IV. Section V shows the simulation results.

II. EXISTING APPROACHES FOR WIDENING TUNING RANGE

Using reverse-biased pn-junctions or MOS capacitance as varactors is the first attempt in widening TR but suffers from high tradeoff between tuning range and PN through AM-PM conversion. Besides, at high frequencies due to device parasitics and limited C_{max}/C_{min} ratio their tuning range is compromised greatly. Their continuous nature, however, favors extensive use in combination with other approaches. Continuous electrical tuning through coupling strength variation in cross-coupled QVCO has been used in [3]. Though having large tuning range their PN degradation is considerable due to significant amounts of frequency shift from resonance. Transformer based coupling is used in [4] but such impedance transformation is associated with degradation of equivalent Q-factor and requires high power to get reasonable PN. In a BiCMOS realization [5] internal buffers were used to isolate and limit device parasitic capacitances. Poor g_m/I_D in CMOS will lead to large power penalty in using such buffers. Digitally controlled capacitor banks and switched inductors [6] have also been proposed as means for discrete tuning but parasitic effects, loss and non-linearity, of the poor switches especially in low voltage CMOS limit their performance. Here, we explore a technique to extend the tuning range of QVCO using an observation particularly applicable for cross-coupled cores without inadvertently worsening any other performance metric.

III. PROPOSED TECHNIQUE

The basic structure of the QVCO is similar to that reported in [7] and we use the fact that such implementation leads to a frequency of oscillation off-resonance. The coupling strength between the two cores and Q of tank determine the frequency shift away from resonance. Based on the polarity of above-mentioned frequency shift, this type of QVCO has two possible modes of operation. Placing phase shifters in the coupling path, as shown in Fig. 1, produces higher loop gain

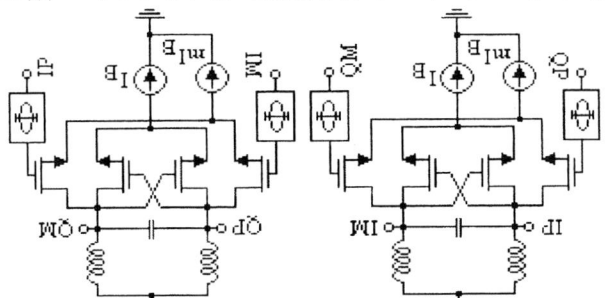

Fig. 1. Conventional coupled LC-QVCO with dedicated phase shifters

in one of the two modes during startup and leads to strong preference in moving towards this mode [8]. The highest loop gain mode in such a system with multiple stable modes eventually becomes the dominant mode of oscillation [9]. We will later show through analysis that the startup of non-dominant mode can be totally inhibited through proper bias and sizing choice.

A simple way of introducing phase shift is through reactive degeneration in the coupling transistors, as shown in Fig. 2. Under assumption of small signal modeling, as explained in Section IV, such degeneration produces a positive or negative phase components in currents contributed by the coupling transconductors with respect to their exciting voltages. Thus, by choosing the type of degeneration, inductive or capacitive, we can steer between two possible frequencies of oscillation. For capacitive degeneration we have a low f_{sr}, low area multi-turn stacked inductor [10] in parallel to maintain bias continuity. Since passive degeneration elements are ideally lossless they do not add noise or power loss as does any dedicated RC phase shifter. Flicker noise problem is alleviated through the use of PMOS transistors both for cross-coupled oscillator core and coupling transistors [11]. The selection of mode is done by switching on or or off appropriate tail current source. Due to the phase shift introduced by the degeneration element we increase sensitivity of quadrature accuracy to IQ tank resonance frequency mismatch. However, it is worth mentioning that under constraints of similar mismatch, power and PN the suggested method shows TR improvement.

The current through coupling pair and negative resistance transistors are kept equal to have good quadrature accuracy in presence of mismatch. With the given power budget, sizing of the negative resistance pair is done to have proper startup with margin and upper-limited to prevent too high parasitic capacitance. Coupling transistors in each path are kept half above size. Under this condition coupling transistors impose significantly lower capacitance than the negative resistance core because of reduced Miller effect and subthreshold region of operation of the inactive path. The degeneration inductor magnitude is upper limited by finite f_{sr}, IQ mismatch and lower bounded by PN degradation constraint. Since Q of the degeneration inductor does not degrade PN noticeably, its form factor can be kept small, f_{sr} of bias continuity inductor, in parallel with degenerating capacitor, and PN constraint determine lower and upper limits of the latter.

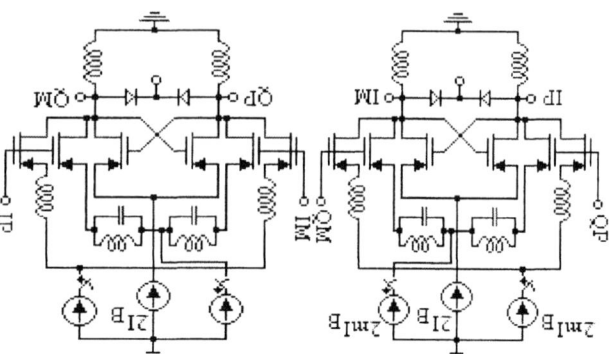

Fig. 2. Proposed QVCO Implementation

IV. ANALYSIS OF DUAL MODE OF OPERATION

In what follows, is given an analysis corroborating our physical understanding of the mechanisms involved in dual mode of operation. Under small signal approximation, valid during startup, the phase shift, alluded to in Section III, can be computed from effective transconductance of coupling transistor given by

$$g_{mc}(s) = \frac{g_{mc,n\,deg}}{1 + (g_{mc,n\,deg} + sC_{gs})Z_{deg}(s)} \quad (1)$$

$Z_{deg}(s)$, the degeneration impedance, is sL for inductive and $1/sC$ for capacitive cases and $g_{mc,ndeg}$ is the non-degenerated transconductance. For values of electrical parameters under consideration we can assume $Z_{deg}(s) \ll 1/sC_{gs}$ and then (1) can be simplified as

$$g_{mc}(s) = \frac{g_{mc,n\,deg}}{1 + sg_{mc,n\,deg}L} \text{ or } \frac{g_{mc,n\,deg}}{1 + \frac{g_{mc,n\,deg}}{sC}} \quad (2)$$

Thus for the two cases we obtain positive and negative phase shifts: $\Phi_{cap} = \tan^{-1}(g_{mc,ndeg}/\omega_{osc}C)$ or $\Phi_{ind} = -\tan^{-1}(\omega_{osc}g_{mc,ndeg}L)$.

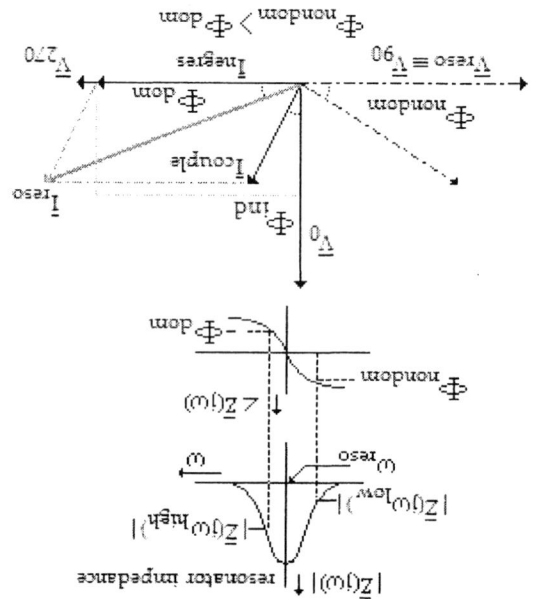

Fig. 4. QVCO VI phasors for Inductively Degenerated Path in Dominant Mode

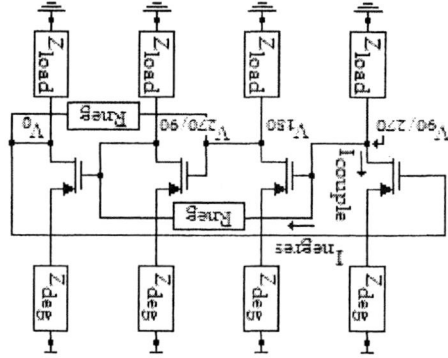

Fig. 3. Simplified Conceptual Diagram of QVCO

V. SIMULATION RESULTS

Simulation results for proposed QVCO in 0.13um CMOS technology are shown in Fig. 5 through 8. The performance is summarized in the numbers below.

between voltage fed to the coupling transistor and current it pumps into concerned tank.

To understand the preferential growth of oscillation into dominant mode, we can look upon the QVCO as a cascade of 4 tuned amplifier stages each producing +/- 90° phase shift to satisfy Barkhausen's phase criterion, as shown in Fig. 3. Phasor diagram for currents and voltages in such QVCO for +90° shifts in inductively degenerated path is shown in Fig. 4. Considering first stage, the current I_{couple} generated by the coupling transistor in response to V_0 adds vectorially with current I_{negres} produced by the negative resistance transistor due to $V_{270/90}$. The resultant I_{reso} flows through resonator impedance $Z(j\omega)$ to generate $V_{reso} = V_{90/270}$. Angle $-\Phi_{dom}$, that V_{reso} subtends with $-I_{reso}$, determines how far off-resonance the oscillation frequency is and controls TR. It is noted that $Z(j\omega) = V_{reso}(j\omega)/-I_{reso}(j\omega)$ due to direction of current chosen. For inductive degeneration we get smaller angle $(-\Phi_{dom})$ and hence higher $|Z(j\omega)| = |Z(j\omega_{high})|$ if the phase shift from one stage to the next is 90° than in the complementary case where $\angle V_{reso}\text{-}I_{reso} = \Phi_{nondom}$. Also, $|I_{reso}/V_0|$ is higher for above phase ordering. A higher $|Z|$ and $|I_{reso}/V_0|$ translate to a larger loop gain $\infty |ZI_{reso}/V_0|^2$ during startup as is derived below. Thus, this mode becomes the dominant mode for oscillation due to exponentially rising amplitude gain over multiple cycles when compared to the other mode.

Representing transconductances of coupling and negative resistance transistors as $g_{mc} = g_{mcr} + jg_{mci}$ and $g_{mn} = g_{mnr}$ respectively, we can show that the gains per differential stage(a), constituted by stages 1 and 3 or 2 and 4 in Fig. 4, for dominant and non-dominant inductive cases are

$$|a_{dom}| = (g_{mnr} - g_{mci})\frac{R_p}{2} - g_{mcr}\beta_{dom}\frac{R_p}{2}\frac{1+\beta_{dom}}{1+\beta_{dom}^2} \quad (3)$$

$$|a_{nondom}| = (g_{mnr} + g_{mci})\frac{R_p}{2} + g_{mcr}\beta_{nondom}\frac{R_p}{2}\frac{1+\beta_{nondom}}{1+\beta_{nondom}^2} \quad (4)$$

$$\text{where, } \beta_{dom} = \frac{g_{mcr}}{g_{mci} - g_{mnr}} \quad (5)$$

$$\beta_{nondom} = \frac{g_{mcr}}{g_{mci} + g_{mnr}} \quad (6)$$

R_p = equivalent parallel resistance of the tanks

$|a|>1 \Rightarrow$ loop gain $\infty |a|^2>1$ and concerned mode can startup.

Since $g_{mci} < 0$, $|a_{dom}/a_{nondom}| > 1$ unconditionally. Moreover, it can be concluded that there exists a range of $\{g_{mc}, g_{mn}\}$ for which the non-dominant mode is not self-starting due to sub-unity loop gain. The frequency of oscillation(ω_{osc}) in inductively degenerated case is higher than nominal resonant frequency(ω_{reso}) of the tank since V_{reso} lags $-I_{reso}$ and it is approximately given by

$$\omega_{osc} = \omega_{high} = \omega_r(1 - \frac{\beta_{dom}}{2Q_{reso}}) \quad (7)$$

β_{dom} being negative for such choice. We can clearly infer that in the capacitive case, θ_{cap} being negative, the behavior will be opposite with a -90° phase shift and ω_{osc} will be $< \omega_{reso}$.

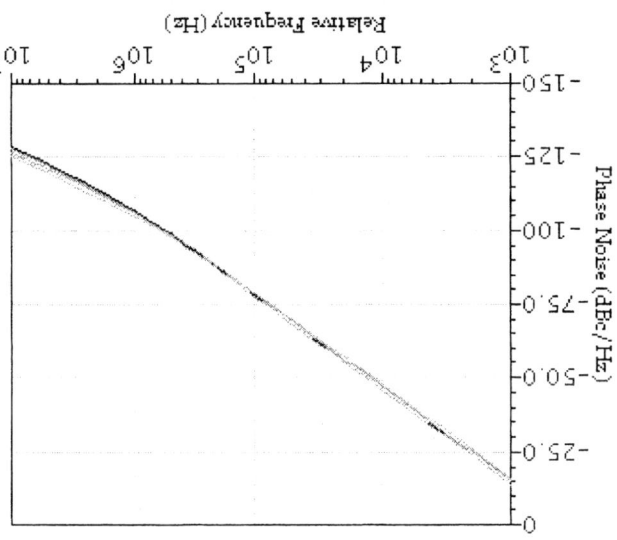

Fig. 5. Phase Noise of the proposed QVCO

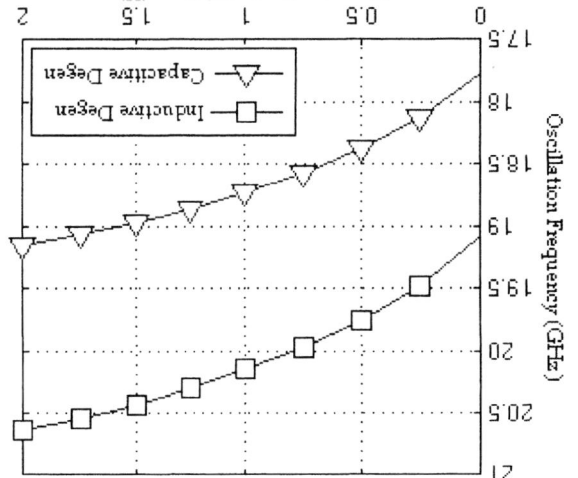

Fig. 6. Tuning range of the QVCO with mode switching and varactor control

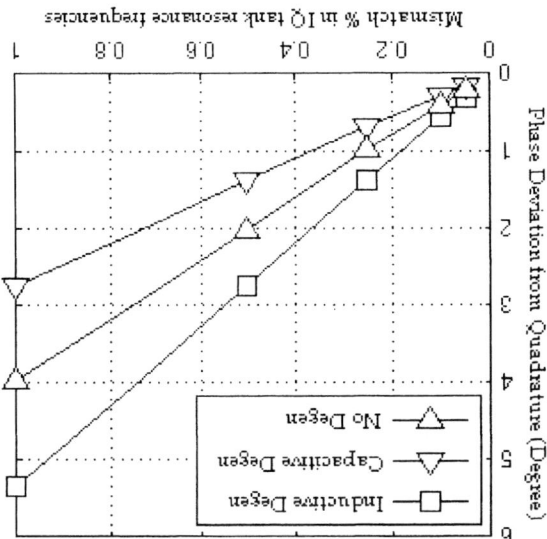

Fig. 7. IQ Mismatch sensitivity

The PN plots are shown in Fig. 5 for extremes of the varactor tuning voltage, 0 and 2V, for the inductive and capacitive degenerations. We can see from the plots that it is very tightly bounded across entire tuning range at constant power due to the symmetric impedance characteristic of resonant load and low PN degradation from small varactors. Transient simulation shows, as per theory, the phase reversal in going from capacitive to inductive degeneration. The tuning range plots validate our assertion on direction of frequency shift. Our choice of device size and bias currents lead to a nominal corner $|a_{dom}/a_{nondom}|\sim1.35$ and $|a_{dom}|\sim1.2$. Thus, only dominant mode is allowed to startup even with process margins. An implementation in same process with no degeneration and consuming identical current has a tuning range from 19.61-21.25-GHz and a PN of -104.3 to -105.9dBc/Hz with similar mismatch sensitivity. Thus the method increases TR by more than 85% with only minimal area penalty.

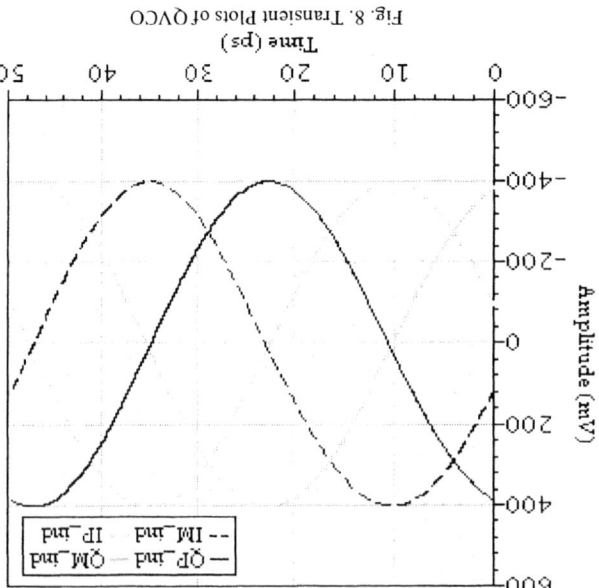

Fig. 8. Transient Plots of QVCO

Supply = 1.2V, Current = 11mA
Phase Noise @ 1M offset = -104.7 to -106.6dBc/Hz
Tuning Range = 17.78-20.64-GHz
Quadrature Inaccuracy at 0.1% f_{reso} mismatch = 0.3-0.55°

TABLE I shows a comparison with other contemporary QVCOs. It is seen that our PN/Power and TR compares favorably with [3], [8] and [12]. [3] has larger fractional TR at higher frequency but consumes significantly higher power. The actual FOM_T for QVCO cores in [3] and [12] will be slightly better, however, since buffer power could not be separated in its calculation.

TABLE I
PERFORMANCE OF RECENT HIGH FREQUENCY QVCOs†

	This Work†	[12]	[8]	[3]
PN@offset(dBc/Hz)	-105.7@1M	-97@1M	-113@2M	-83.5@3M
Tuning Range(GHz)	17.8-20.6	30.6-32.6	4.9-5.23	38-43
Current(mA)	11	43	7.85	46
Power(mW)	13.2	215*	21.2	69*
Process L_{min}(um)	0.13	0.25	-	0.09
NMOS f_T(GHz)	90	70	35	150
FOM_T(dBc/Hz)	-183.3	-161.5	-164.1	-149.5

†PN and Power were averaged over TR where data was available + Simulated
*With buffers

VI. CONCLUSION

A new technique to extend TR of high frequency QVCO has been investigated. It exploits the existence of two possible modes of oscillation of cross-coupled core based quadrature oscillators. Reactive degeneration in coupling transistors help in switching between the modes. Simulation results for a 0.13um CMOS implementation show a 17.78 to 20.64-GHz TR and close in PN of -104.4 to -106.7dBc/Hz at 1MHz offset from center frequency. The vco consumes 11mA from 1.2V supply. Improvement in TR comes with minimal area impact and does not worsen any critical performance metric.

REFERENCES

[1] B. Razavi, *RF Microelectronics*, 1st ed. Upper Saddle River, NJ: Prentice-Hall, 1998.

[2] M. Meghelli, A.V. Rylyakov, S.J. Zier, M. Sorna and D. Friedman, "A 0.18 um SiGe BiCMOS receiver and transmitter chipset for SONET OC-768 transmission systems", in *IEEE ISSCC Dig. Tech. Papers*, 2003, pp. 230 - 231.

[3] F. Ellinger and H. Jackel, "38-43 GHz quadrature VCO on 90 nm VLSI CMOS with feedback frequency tuning", in *IEEE MTT-S Int. Microwave Symposium Dig. Tech. Papers*, 2005, pp. 1701-1703.

[4] K. Kwok, J. R. Long and J. J. Pekarik, "A 23-to-29GHz Differentially Tuned Varactorless VCO in 0.13um CMOS", in *IEEE ISSCC Dig. Tech. Papers*, 2007, pp. 194 - 195.

[5] B. Jung and R. Harjani, "High-frequency LC VCO design using capacitive degeneration", *IEEE J. Solid-State Circuits*, vol. 39, no. 12, pp. 2359 - 2370, Dec. 2004.

[6] S. Yim and K. K. O, "Switched resonators and their applications in a dual-band monolithic CMOS LC-tuned VCO", in *IEEE Trans. Microw. Theory and Tech.*, vol. 54, no. 1, pp. 74 - 81, Jan. 2006.

[7] A. Rofougaran, J. Rael, M. Rofougaran and A. Abidi, "A 900 MHz CMOS LC-oscillator with quadrature outputs", in *IEEE ISSCC Dig. Tech. Papers*, 1996, pp. 392 - 393.

[8] P. van de Ven, J. van der Tang, D. Kasperkovitz and A. van Roermund, "An optimally coupled 5 GHz quadrature LC oscillator", in *Symp. VLSI Circuits Dig. Tech. Papers*, 2001, pp. 115 - 118.

[9] T. Piessens and M. Steyaert, "Oscillator pulling and synchronisation issues in self-oscillating class D power amplifiers", in *Proc. ESSCIRC*, 2003, pp. 529 - 532.

[10] A. Zolfaghari, A. Chan and B. Razavi, "Stacked inductors and transformers in CMOS technology", *IEEE J. Solid-State Circuits*, vol. 36, no. 4, pp. 620 - 628, April 2001.

[11] J. Kim, O. Plouchart, "The impact of device type and sizing on phase noise mechanisms", *IEEE J. Solid-State Circuits*, vol. 40, no. 2, pp. 360 - 369, Feb. 2005.

[12] W. L. Chan, H. Veenstra and J. R. Long, "A 32 GHz quadrature LC-VCO in 0.25 um SiGe BiCMOS technology", in *IEEE ISSCC Dig. Tech. Papers*, 2005, pp. 538 - 616.

[13] M. Tabll, J. Safran, A. Ray and L. Wagner, " A power-optimized widely-tunable 5-GHz monolithic VCO in digital SOI CMOS technology on high resistivity substrate", in *Proc. Int. Symp. Low Power Electronics and Design*, Aug. 2003, pp. 434-439.

Top-Down Simulation Methodology of a Mixed-Signal Read Channel Using Standard VHDL

R. Bogdan Staszewski

Texas Instruments, Dallas, TX 75243, USA

Abstract—This paper presents a mixed-signal system modeling and simulation methodology using an event-driven simulator that supports real-valued signals, such as standard VHDL. Success of this methodology has been demonstrated by a commercial 550 MHz **Partial Response Maximum Likelihood (PRML)** magnetic recording read channel. The read channel IC is of mixed-signal design type with 30% analog and 70% digital content. The digital portion has been synthesized from the RTL subset of VHDL (1987 standard). The analog part has been behaviorally modeled using the 1993 standard version of VHDL. Five abstraction levels of digital circuits modeling are also described.

I. INTRODUCTION

The modern synchronous magnetic recording system [1] [2] consists of the read, write and servo blocks (Fig. 1). The write path takes a byte-wide digital data, encodes it into a (d, k) run-length-limited (RLL) code, then scrambles or randomizes the data, pre-compensates the transition edges for the non-linear bit shift of the magnetic media, and writes the data onto the disk media. The servo block keeps track of the location of the data on the media, controls the position of the drive heads, and adjusts the motor speeds. The servo mechanism must also be able to handle wide temperature variations and severe vibrations especially in disk units installed in portable systems. The read path is responsible for taking a low-amplitude (1 mV$_{pp}$) analog readback signal from the magnetic media, amplify, sample, equalize and decode it, such that the resulting bit error rate would be on the order of 10^{-6}. The error correction circuit (ECC) is then used to reduce the user-data byte error rate to about 10^{-12}. The typical blocks within the read path are automatic gain control (AGC), continuous-time filter (CTF), A/D converter (ADC), digital finite impulse response filter (FIR), Viterbi sequence detector, decoder and de-scrambler. In addition, the timing recovery loop extracts

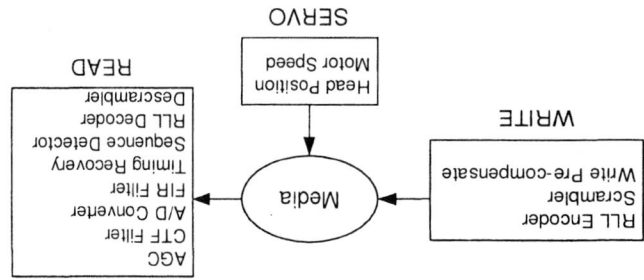

Fig. 1. Magnetic recording disk drive system blocks.

the bit-rate clock from the sampled partial response maximum likelihood (PRML) data.

The key advantage of PRML is that it works with sequences of received data, rather than single bits, comparing data sequences and determining which one was most likely to have caused the observed signal. It is therefore more tolerant of inter-symbol interference (ISI) than the pulse peak detection. Basically, the idea behind partial response (PR) signaling is to increase the density so that ISI occurs, but control it so that the detector knows what signal shape to expect. This read channel supports both PR4 $(1-D^2)$ polynomial with three data levels, where D is a one-symbol delay operator) and EPR4 $(1+D-D^2-D^3$ polynomial with five data levels) pulse responses.

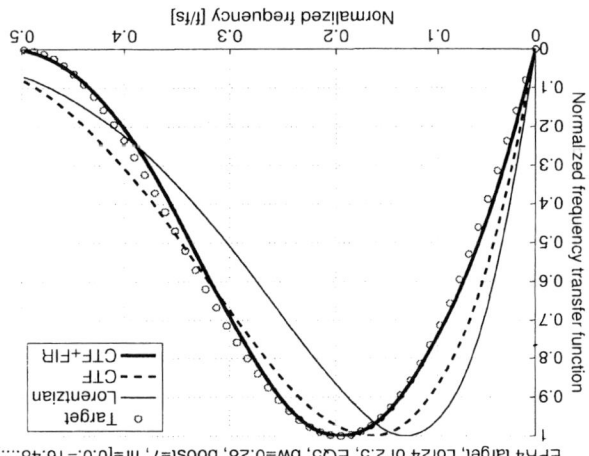

Fig. 2. EPR4 equalization sharing between CTF and FIR filters.

Fig. 2 shows the EPR4 equalization process shared by both a continuous-time filter (equiripple type with bandwidth of $f_c = 0.28$ and boost of 7 dB) and an 8-tap FIR filter [3]. The incoming signal (leftmost line) is modeled as a Lorentzian response of 2.4-th power. The CTF filter performs a partial equalization (dotted line) and the FIR filter completes it (rightmost line) such that is approximately fits the EPR4 target (circles). Fig. 3 shows the actual ADC and FIR outputs (in time domain) under some extreme operating conditions.

There have been a variety of read channel or, more generally, communication channel modeling methods. At the pure system level there are C and MATLAB models, which are highly abstract and with very weak links to the actual hardware. On the opposite side of the spectrum, the system could be modeled at a very low level entirely in SPICE for analog intensive systems or in SPICE and Verilog (or VHDL) combination, with a varying degree of link between the two disparate simulation engines (e.g., TI-SPICE and Verilog co-simulation backplane). Establishing a link to a non-event-driven engine, such as SPICE, results in a hefty price of a simulation performance, thus making it impossible to determine the very basic figure of merit of a communication channel, such as bit error rate.

Previous implementations of read channels, especially analog intensive with a low digital contents, have been successfully modeled by the author with MAST, a proprietary (product of Analogy) mixed-signal modeling language running Saber simulator [4]. Saber, despite integrating analog and digital simulators in a single engine, suffers from weak support for digital modeling constructs, such as lack of buses, making it unsuitable for more complex digital systems. In addition, the digital subset of MAST is not synthesizable.

We demonstrate a system and a simulation environment that are based on a standard single-core VHDL simulator. This system is well suited for digitally intensive applications with a fair amount of analog circuit content. Extensive links to a file system for pre- and post-processing facilitate rich simulation and analysis environment. The main advantage of the single simulation engine is that it allows a seamless integration of all hardware abstraction levels in a uniform environment. The single most important feature of a standard VHDL language, which makes it far superior than Verilog for mixed-signal designs is its support of real or floating point type signal. Widespread simulation and synthesis support of

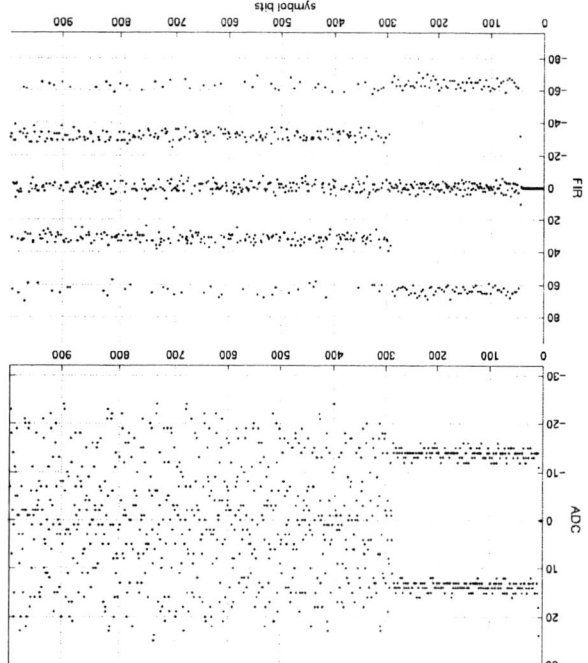

Fig. 3. Unequalized ADC and equalized FIR outputs.

a standard VHDL language makes it possible for a complex communication system to achieve "build what we simulate, and simulate what we build" goal.

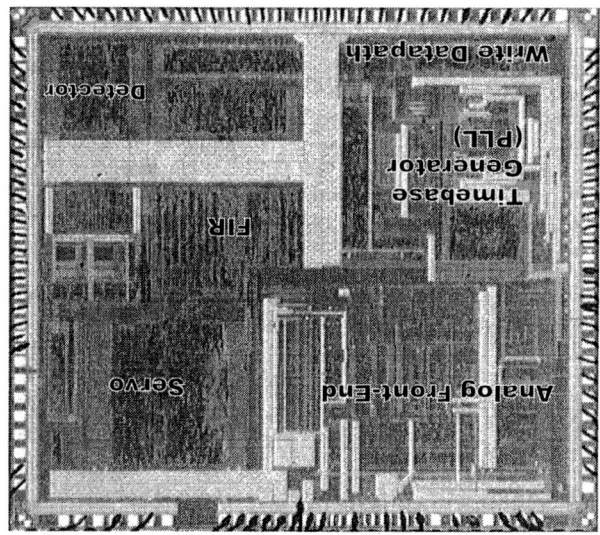

Fig. 4. Micrograph of the read channel IC.

The presented techniques have been validated and successfully used in design, modeling, simulation, and verification of a production 550 MHz PRML magnetic recording read channel IC in 0.18 μm L-effective CMOS technology. The chip is of mixed-signal design type with about 30% analog and 70% digital content and comprises about 14 mm^2 silicon area. Fig. 4 shows the chip micrograph.

II. ANALOG CIRCUITS

The analog circuits are usually described behaviorally using s-domain (continuous-time and continuous-amplitude domains) equations. In order to be able to employ an event-driven simulator, a conversion to a z-domain (sampled-time and continuous-amplitude domains) was performed. Fixed-time steps with fine resolution were selected. It is easy to show that the oversampling ratio of about 10 is quite practical and sufficient for read channel applications.

A continuous-time 5-th order equiripple (Chebyshev) filter has been modeled using the impulse invariance method. The essential idea in the design of impulse response filters is that the unit-response of the sampled filter be equal to the sampled values of the impulse response of the desired analog filter. The discrete approximation of the frequency response of this method is very good for higher values of oversampling ratio (OSR). The approximation error solely depends on the amount of aliasing. For example, the Nyquist frequency for OSR=10 would be five times higher than the sampling frequency, or 20 times higher than the 3 dB filter cutoff frequency of a typical CTF filter found in read channel IC chips. Thus, considering sharp roll-off of the filter's transition band, the aliasing amount and, therefore, the approximation error are negligible.

The 5-th order equiripple filter is architecturally implemented as an IIR filter whose transfer function is rational with 5-th order polynomial in the denominator and 4-th order polynomial in the numerator. Since the continuous-time filter settings, such as cutoff frequency, boost and zeros asymmetry, are dynamically changed during the system operation, it is necessary to implement a mechanism by which the coefficients of the transformed sampled IIR filter are updated. This could be done either by real-time recalculation or by a table lookup. We have chosen the latter method where the coefficients are conveniently pre-computed once before the system is built and then stored in a file. Whenever the CTF filter parameters are changed during the simulation runs, the file would be accessed for IIR coefficient look-up. The alternative method of coefficient recalculation is ill suited for VHDL due to the sheer mathematical complexity that involves finding residues and poles of the partial-fraction expansion of the transfer function. It is easy to show that the other competing method of the analog filter approximation problem using sampled filters, the bilinear transform method, is not well suited for equalizing filters, such as ours, because of the nonlinear frequency warping.

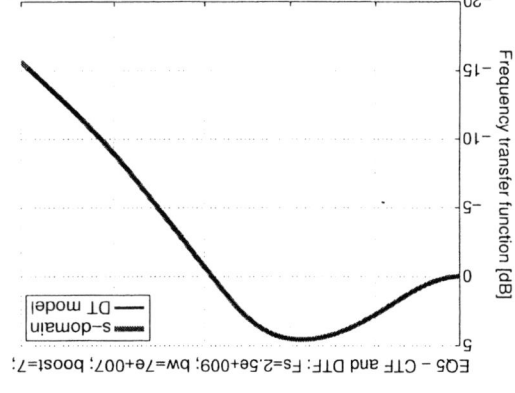

EQ5 - CTF and DTF; Fs=2.5e+009; bw=7e+007; boost=7;

Fig. 5. Transfer function of the CTF.

Since VHDL does not support frequency-domain analysis, a dedicated VHDL testbench to extract the frequency-domain transfer function characteristic was written that performed frequency sweep of time-domain simulations. The result is shown in Fig. 5.

Any analog filter approximation implementation requires that the fine samples be equally spaced in order to avoid unstable frequency reference and associated frequency modulation. Essential filter parameters, such as 3 dB cutoff frequency, f_c, are originally expressed in frequency units of Hz. In the discrete domain, the θ_c cutoff frequency is expressed as: $\theta_c = 2\pi f_c(t_0/2)$. This equation shows that the non-uniform sampling (variation of t_0) will modulate the filter parameter. Due to this very reason, the entire analog section was implemented "clock-less" (an oversampled clock signal is of non-physical nature and does not fit well with the overall strategy of pin compatibility at different abstraction levels) but driven from a stream source that generates fine equally-

spaced analog samples. The "analog" data signal contains not only sample values but also carries the event activity.

The analog section follows a unidirectional signal flow in s-domain. At each successive block, care must be taken to only propagate incoming events and not to generate any new ones. This is usually accomplished by making the incoming data input the only entry in the process sensitivity list that generates the output data, as shown in the listing below of a programmable-gain amplifier (PGA).

```
1  entity PGA_LA is
2    port (
3      DTI    : in real;
4      GAIN   : in real := 0.0;
5      DTO    : out real := 0.0
6    );
7  end entity PGA_LA;
8
9  architecture BEHAV of PGA_LA is
10   signal GAIN_DB, GAIN_X : real;
11 begin
12   GAIN_DB <= 10.0 * (log10((1.0 + (GAIN / 17.0))
13     / (1.0 - (GAIN / 17.0)))) - 2.0;
14   GAIN_X  <= 10.0 ** (GAIN_DB / 20.0);
15   process (DTI) is
16   begin
17     DTO <= DTI * GAIN_X;
18   end process;
19 end architecture BEHAV;
```

Certain functions, such as CTF or an interpolator (see Sec. III), require not only the input events, but also the signal activity, to trigger the process. The latter will guarantee the trigger at fixed intervals, even though the signal amplitude might not change between two consecutive samples, as in noiseless case.

III. INTERPOLATOR

A magnetic recording read channel, being an example of a synchronous communications system, features a clock extraction circuitry whose purpose is to synchronize an internally generated clock to the incoming data rate. Therefore, the two principal clock domains, the oversampled "analog" domain and the bit-clock digital domain, are *totally asynchronous* to one another. This requires an interpolator that serves as an interface between the analog/continuous-time and clock-sampled/digital domains. A first order, linear interpolator was chosen. It is easy to show that for higher value of oversampling ratio the second, or higher, order interpolators are not worth an extra computational burden and implementational complexity. The VHDL code of the interpolator is listed below:

```
1  architecture BEHAV of IPL_LA is
2  --update OS memory or sample
3  type STATE_IPL_TYPE is (UPD_OS, UPD_CLK);
4  begin
5  OS_OR_CLK: process (DTI, CLK) is
6  -- time of previous input activity
7  variable TIME_1: time := 0 ns;
8  -- time of last clock event
9  variable TIME_C: time := 0 ns;
10 -- value at previous input activity
11 variable DTI_1: real := 0.0;
12 -- interpolated value at last clock event
13 variable DTO_C: real := 0.0;
14 variable STATE_IPL: STATE_IPL_TYPE := UPD_OS;
```

```
15  begin
16    if now = 0 ps then
17      ... Initialize variables to zero
18    end if;
19    if DTI'active then
20      case STATE_IPL is
21        when UPD_OS =>
22          DTI_1 := DTI;
23          TIME_1 := now;
24        when UPD_CLK =>
25          DTO_C := DTI_1 + (DTI-DTI_1)*
26            T2R(TIME_C-TIME_1)/T2R(now-TIME_1);
27          DTI_1 := DTI;
28          TIME_1 := now;
29          STATE_IPL := UPD_OS;
30      end case;
31    end if;
32    if CLK'event and CLK = '1' then
33      DTO <= DTO_C;
34      TIME_C := now;
35      STATE_IPL := UPD_CLK;
36    end if;
37  end process;
38 end architecture BEHAV;
```

IV. DIGITAL BLOCKS

TABLE I
VHDL ABSTRACTION LEVELS.

Level 1	Mathematical equations and high-level behavioral description. Parameterized for easy analytic "what-if" questions. Optimized for simulation speed and flexibility. Fast enough to replace MATLAB for a bit error rate analysis. Includes important hardware-related non-idealities and second-order effects. .l1.vhd file suffix.
Level 2	Mathematical equations in integer domain. Implying the underlying architectural structure. 100% pin-compatible with Level 3. Can be used for top-level connectivity verification. .l2.vhd file suffix.
Level 3	Synthesizable register transfer level (RTL).
Level 4	Gate-level netlist produced by a synthesis tool.
Level 5	The actual gate-level netlist extracted from an auto place and route (APR) tool. Accompanied by the cell and wire timing information (Vital or SDF).

The digital blocks are modeled at various abstraction levels, as shown in Table 1. As an example, Level 1 representation of an FIR might describe the behavior with a simple direct-form finite impulse response filter equation using real numbers for input, output, coefficients and all the intermediate signals. Level 2 representation of the actual transpose FIR implementation might show the top-level structure of major building blocks, which are then modeled behaviorally on the bit level using integers. Second order effects, such as LSB truncation and rounding as well as MSB clipping, are included. Level 3 representation has the same I/O behavior as Level 2, but its RTL representation drives the synthesis of gate connectivity. Each level of modeling is expected to improve the simulation time by an order of magnitude. For example, the Level 1 complete read path model processes 400 cycles per second of simulation, whereas Level 3 processes 20 cycles per second. This unified approach ensures interoperability of various abstraction levels within a single simulation environment. This allows simulating a system with a mixture of synthesized blocks and those that are still at mathematical description levels.

V. SUPPORT OF DIGITAL STREAM PROCESSING

The magnetic read/write channel can be viewed as another form of a classical digital communications channel where the write path is the transmitter, the magnetic media and head combination is the transmission medium, and the read path is the receiver. Digital communications channel evaluation usually requires processing of a long stream of digital data. Some measurements, such as bit error rate measurements, require as much as tens of millions of data bits, therefore, efficient and fast algorithms to store, retrieve and access the data are necessary. Throughout this system, a time-causal operation and a linear complexity order is forced on the processing algorithms. As a result, the temporary storage requirement (RAM or disk swap space) is constant and the simulation time is only linearly proportional to the digital stream length.

File storage of digital data stream supports a character-based file format where each uncommented ASCII character '0', '1', 'U', 'X', '-' corresponds to logic low, high, unknown, contention and don't care, respectively. Comments are UNIX-shell like and start with the pound character '#' and run until end-of-line.

VI. SIMULATION TIME PERFORMANCE COMPARISON

Mixed-signal simulation performance of a digitally-intensive read channel system, when modeled in a standard event-driven VHDL, is far superior (and much more efficient) than any dual-engine co-simulator or even an integrated engine simulator, such as Saber, which carries performance burden of a true continuous-time simulator core. A similar system implemented in Saber/MAST was found to run two-to-three orders of magnitude slower.

VII. CONCLUSION

This paper presented a mixed-signal system simulation methodology of a commercial PRML magnetic recording read channel IC chip. Numerous problems of modeling continuous-time analog circuits using a standard VHDL event-driven simulator have been solved. Single-core simulation system supports an integrated environment with generation of stimulus and post-processing the data. VHDL modeling abstraction levels have been proposed. Support of a digital stream processing has also been described.

REFERENCES

[1] H. Kobayashi and D. Tang, "Application of partial-response channel coding to magnetic recording systems," IBM J. Res. Develop., vol. 14, pp. 386-375, July 1970.

[2] R. Cideciyan et al., "A PRML system for digital magnetic recording," IEEE Journal on Selected Areas in Communication, vol 10, pp. 38-56, Jan. 1992.

[3] R. B. Staszewski, K. Muhammad, and P. T. Balsara, "A constrained asymmetry LMS algorithm for PRML disk drive read channels," IEEE Trans. on Circuits and Systems II, vol. 48, pp. 793-798, Aug. 2001.

[4] A. Mantooth, "Modeling with an Analog Description Language," Kluwer Academic Publishers.

Modeling of an Electronic Noise and Media in a Magnetic Recording Read Channel Using VHDL

R. Bogdan Staszewski

Texas Instruments, Dallas, TX 75243, USA

Abstract—This paper presents a modeling and simulation methodology of two challenging aspects of a partial response maximum likelihood (PRML) magnetic recording read channel: electronic noise and magnetic media. The methodology is based on an event-driven simulator that supports real-valued signals, such as standard VHDL. A few examples are presented for the noise source, PLL, jitter/wander modeling, and magnetic media waveform creation. Success of this methodology has been demonstrated through a commercial 550 MHz PRML read channel IC. The read channel IC is of a mixed-signal design type with 30% analog and 70% digital content.

I. INTRODUCTION

Todays hard-disk drive electronics employ sophisticated signal processing methods to meet the demand of continually increasing storage capacity and transfer data-rate requirements. One popular approach is class IV partial (or extended partial) response with maximum likelihood detection (PRML) [1] [2]. Economical implementation of PRML read channels has been demonstrated using analog signal processing [3]. Fig. 1 shows a typical disk drive PRML read channel. As shown, the preamp is used to amplify small signals read from the head and is typically situated close to the head-assembly unit. The signal is then passed to the variable gain amplifier (VGA) stage which adjusts the signal amplitude for further processing. Next, the signal is passed through the continuous-time filter (CTF) which limits the noise bandwidth and provides anti-alias filtering prior to sampling the signal. The equalizer provides high frequency boost in order to match PRML spectrum. It is desirable to have an adaptive equalizer to remove time dependent channel variation. This provides the system ability to adapt to the wide variabiliy of magnetic media, read heads, front-end analog electronics and environmental factors. The use of magneto-resistive (MR) heads is currently dominating their thin-film inductive head counterparts, mainly, due to their

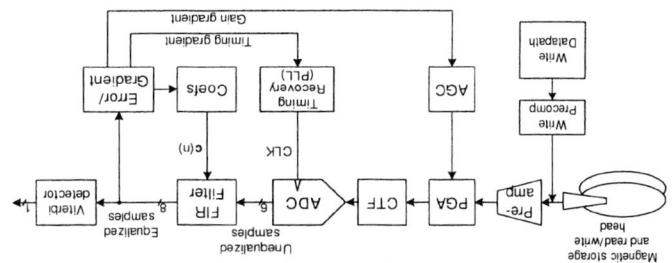

Fig. 1. Magnetic recording disk drive system.

lower cost. However, their transfer function response is non-symmetric for positive and negative pulses and exhibits a larger amount of nonlinearity. Consequently, the equalizer's adaptive algorithm must conform to more stringent standards of performance as it must be able to correctly shape the read-back data to the target PRML polynomial despite all these non-linear imperfections. Finally the equalizer output is passed to the sequence detector to recover the original data. Traditionally, adaptive equalizers have been implemented in digital domain, requiring analog-to-digital converter (ADC).

The sheer complexity of the read channel electronics and the associated magnetic media makes it imperative to model and simulate the complete system prior to tape-out [4]. There have been a variety of read channel or, more generally, communication channel modeling methods. At the pure system level there are C and MATLAB models, which are highly abstract and with very weak links to the actual hardware. On the opposite side of the spectrum, the system could be modeled at a very low level entirely in SPICE for analog intensive systems or in SPICE and Verilog (or VHDL) combination, with a varying degree of link between the two disparate simulation engines (e.g., TI-SPICE and Verilog co-simulation backplane). Establishing a link to a non-event-driven engine, such as SPICE, results in a hefty price of a simulation performance, thus making it impossible to determine the very basic figure of merit of a communication channel, such as bit error rate.

This paper presents novel modeling and simulation techniques of the electronic noise and magnetic media.

The presented techniques have been validated and success-fully used in design, modeling, simulation, and verification of a production 550 MHz PRML magnetic recording read channel IC in 0.18 μm L-effective CMOS technology. The chip is of mixed-signal design type with about 30% analog and 70% digital content and comprises about 14 mm^2 silicon area. Fig. 2 shows the chip micrograph.

II. RANDOM NUMBER GENERATOR

A pseudo-random number generator is needed to create a long stream of the digital input stimulus, model an electronic and media colored input noise, add jitter and wander deviations to the derived system clock and add transition shift to the magnetic domains. It became clear, from the very beginning, that system-supplied pseudo-random number generators do not usually have good random properties. A typical imple-mentation of *rand()* ANSI C function call or *uniform()* procedure call of the IEEE math-real VHDL package uses the

Fig. 2. Micrograph of the read channel IC.

linear congruential method which, while being very efficient and fast, suffers from sequential correlation on successive calls. There is a danger that using it might skew evaluated performance of a communication system by, for example, not exercising all possible paths through a Viterbi detector. A good uniformly distributed random number generator with virtually no sequential correlation is shown below based on Park and Miller algorithm with Bays-Durham shuffle [5].

```
1   process (CLK, ENA) is
2   variable S1: integer := SEED1;
3   variable S2: integer := SEED2;
4   variable VAL: real;
5   variable INITIAL: boolean := true;
6   variable V: REAL_VECTOR (1 to 97);
7   variable Y: real;
8   variable J: integer;
9   begin
10  if CLK'event and CLK='1' and ENA='1' then
11  if INITIAL then
12  -- one-time initialization
13  for INDEX in 1 to 97 loop
14  -- load the shuffle table
15  uniform(S1, S2, VAL);
16  V(INDEX) := VAL;
17  end loop;
18  -- initial random number
19  uniform(S1, S2, VAL);
20  Y := VAL;
21  INITIAL := false;
22  end if;
23  -- start here when not initializing
24  -- Use previously saved random number,
25  -- Y, to get an index, J, between 1 and 97.
26  -- Then use the corresponding V(J) for
27  -- both the output and the next J.
28  J := 1 + integer(96.0 * Y);
29  -- range error checking
30  assert (J >= 1 and J <= 97)
31  report "Index J out of bounds!"
32  severity failure;
33  Y := V(J);
34  uniform(S1, S2, VAL);
35  V(J) := VAL;
36  -- map uniform random deviate into
37  -- logic '0'/'1' space
38  if Y < RATIO_1_0 then
39  DTO <= '1' after TD;
40  else
41  DTO <= '0' after TD;
42  end if;
43  elsif ENA'event and ENA='0' then    -- reset
44  DTO <= '0' after TD;
45  end if;
46  end process;
```

A random number generator with gaussian (normal) distribution was built (and shown below) based on the Box-Muller method [5].

```
1   architecture BEHAV of SRC_NOISE is
2   signal SMP: bit := '0';
3   begin
4   process (SMP) is
5   variable S1: integer := SEED1;
6   variable S2: integer := SEED2;
7   variable ISET: boolean := false;
8   variable FAC, RSQ, V1, V2, GSET: real;
9   begin
10  if not ISET then
11  GET_RSQ: loop
12  uniform(S1, S2, V1);
13  V1 := 2.0 * V1 - 1.0;
14  uniform(S1, S2, V2);
15  V2 := 2.0 * V2 - 1.0;
16  RSQ := V1*V1 + V2*V2;
17  exit when RSQ <= 1.0;
18  end loop GET_RSQ;
19  FAC := NRMS * SQRT(-2.0 * LOG(RSQ) / RSQ);
20  GSET := V1 * FAC;
21  DTO <= V2 * FAC;
22  ISET := true;
23  else
24  DTO <= GSET;
25  ISET := false;
26  end if;
27  end process;
28  GEN_CLOCK: process (SMP) is
29  begin
30  SMP <= not SMP after TIME_STEP;
31  end process GEN_CLOCK;
32  end architecture BEHAV;
```

III. CLOCK JITTER AND WANDER EFFECTS

We have successfully modeled jitter and wander sources of an internal phase lock loop (PLL) circuit. From the system's perspective it is important to differentiate between the two contributors to uncertainty of the clock ideal sampling instances. The wander is a slowly varying component of the PLL phase error and is modeled as random walk, whereas the jitter is that fast-varying phase component which regresses to the mean.

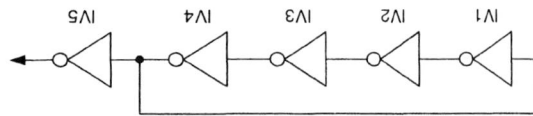

Fig. 3. 4-stage ring oscillator.

Fig. 3 shows a simplified schematic of a 4-stage differential ring oscillator commonly used in high frequency PLL circuits.

Its operating frequency and significant time instances (i.e., time stamps of the output clock rising edges) are determined by the sum of the four inverter delays IV1 through IV4:

$$T_0 = \sum_{i=1}^{4}(t_i + \Delta t_i) \tag{1}$$

where, T_0 – instantaneous time period, t_i – nominal delay of the ith inverter, Δt_i – instantaneous random deviation of the ith inverter.

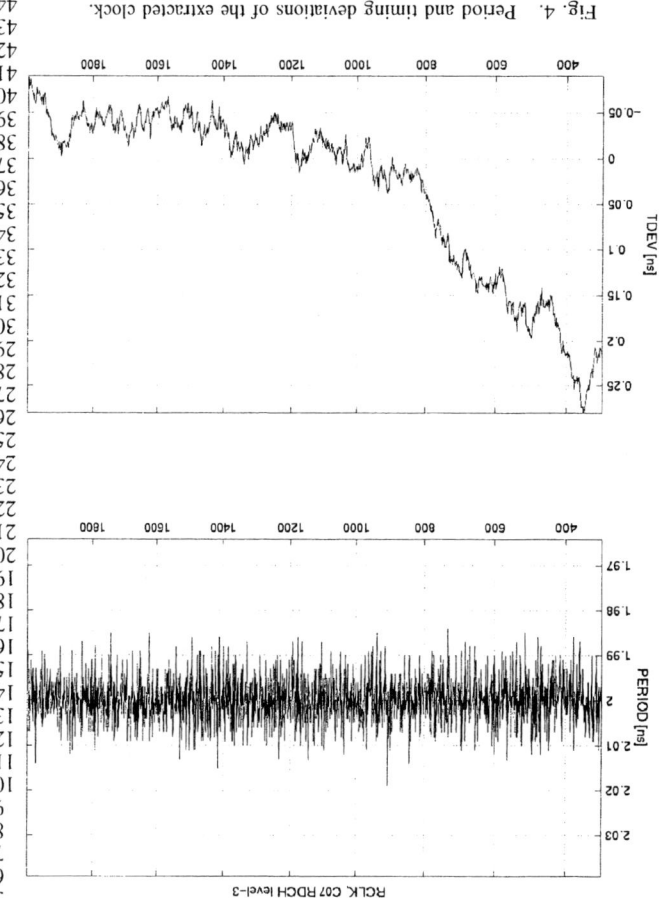

Fig. 4. Period and timing deviations of the extracted clock.

The fifth inverter, IV5, is outside of the feedback loop and is used as a buffer. Except for the effects of its loading input, it does not effect the operational frequency. A timing disturbance on any of the feedback loop inverters IV1-4 will cause the output phase to get accumulated and, if not corrected through other means, will not show any regression to the mean. This phase disturbance behavior fits the Markov chain random walk model [6] and is referred to as wander. The inverter jitter, since its phase disturbance effects on the time interval error are not cumulative. Fig. 4 shows the instantaneous period deviation and wander, expressed as time interval error or timing deviation (TDEV), of the system.

IV. MAGNETIC MEDIA

Magnetic media was modeled using a waveform synthesizer and includes parametric description of channel density, media noise, jitter/wander, and non-linear bit shift effects. It produces an oversampled "analog" readback head signal as a response to the write head magnetization current and is based on the magnetic channel model and disturbance parameters. A listing below shows the media model parameters.

```
 1  entity MEDIA is
 2    generic (
 3      -- channel density: number of symbols per PW50
 4      D: real := 2.0;
 5      -- number of samples per symbol interval
 6      N: integer := 10;
 7      -- number of symbol intervals per
 8      -- transition response on each side
 9      M: integer := 10;
10      -- transition response type:
11      -- 0=PR4, 1=EPR4, 2=Lor2, 4=Lor4, 24=Lor2.4
12      LORTYPE: integer := 24;
13      -- thinning factor of the written pulses
14      THIN: integer := 1;
15      -- rms of the internal AWGN noise generator
16      NRMS: real := 0.0;
17      -- rms of the internal AWGN jitter generator
18      JRMS: time := 0 ps;
19      -- use time-based random number generator seed
20      RANDOMIZE: boolean := true;
21      -- magnetic media waveform output file
22      FILE_NAME: string := "read.rtx"
23    );
24    port (
25      -- input data (write head transition current)
26      DTI: in std_logic;
27      -- control enable
28      ENA: in std_logic;
29      -- symbol interval (clock period) in seconds
30      T: in time;
31      -- amplitude of positive transition response
32      AP: in real;
33      -- amplitude of negative transition response
34      AN: in real;
35      -- pulse offset, normally = 0
36      OFFSET: in real;
37      -- deterministic non-linear bit shift delay [1]
38      -- only next-to-previous symbol in transition
39      SHIFT1: in time;
40      -- only previous symbol in transition
41      SHIFT2: in time;
42      -- two previous symbols in transition
43      SHIFT3: in time;
44    );
45  end entity MEDIA;
```

Distinction is being made between static and dynamically changing parameters. The former include channel density and step response Lorentzian type and are modeled as generics, whereas the latter include magnitude or dc offset and are modeled as signals.

We solved an interesting algorithmic problem of writing a time-causal media model to which the only input is the write data stream without an accompanying clock. The challenge lies in the fact that the data could exhibit a large amount of jitter or the write pre-compensation circuit (WPC) might have non-linearly delayed the phase. As a reminder, supplying the accompanying clock would easily solve the problem, but the extra chip output pin would be non-physical.

on each side and is derived from the pulse template and possibly adjusted for the noise, fractional bit delay (first-order time interpolation), amplitude and dc offset instantaneous variations. The working array, war, shifts and outputs the least recent samples corresponding to the number of bits variations between the transitions and then performs superposition with the current array.

V. READBACK WAVEFORM STREAM

The magnetic media waveform generator produces an ASCII file that stores the oversampled ($N = 10$ in the listing example in Sec. IV) representation of the analog readback signal. Its file format stores only the signal amplitude samples, thus allowing for an easy time shifting and frequency scaling of the readback waveform. During the readback process, the stored samples are read and then noise is added and output to the read path at fixed-time intervals. No information about the exact symbol rate and timing is communicated to the read channel. In fact, it is a function of the timing recovery circuit (see Fig. 1) to facilitate the synchronous bit detection.

A listing in Sec. II shows a VHDL code of a gaussian random number generator used as an AWGN noise source.

VI. CONCLUSION

We have presented a modeling and simulation methodology of two challenging aspects of a partial response maximum likelihood (PRML) magnetic recording read channel: electronic noise and magnetic media. The flow is part of an integrated system design environment based on a standard VHDL hardware description language. The digitally-oriented modeling and simulation environment is compatible with the predominantly digital nature of this mixed-signal IC, which is about 70% digital and 30% analog. We have disclosed efficient VHDL code of additive white gaussian noise (AWGN) with good statistical properties, which is used as a noise source of various electronic circuits, including jitter/wander of PLL oscillators. In addition, we described a magnetic media modeling that is in agreement with physical properties of magnetic domains, such as bit shift due to neighboring domain polarizations.

REFERENCES

[1] H. Kobayashi and D. Tang, "Application of partial-response channel coding to magnetic recording systems," *IBM J. Res. Develop*, vol. 14, pp. 386-375, July 1970.

[2] R. Cideciyan *et al.*, "A PRML system for digital magnetic recording," *IEEE Journal on Selected Areas in Communication*, vol 10, pp. 38-56, Jan. 1992.

[3] S. Kiriaki, T. L. Viswanathan, G. Feygin, et al., "A 160-MHz analog equalizer for magnetic disk read channels," *IEEE Journal of Solid-State Circuits*, vol. 32, no. 11, pp. 1839-1850, Nov. 1997.

[4] R. B. Staszewski, "Top-down simulation methodology of a mixed-signal read channel using standard VHDL," *Proc. of Sixth IEEE Dallas Circuits and Systems Workshop: Design, Application, Integration and Software (DCAS-07)*, Nov. 2007.

[5] W. H. Press, S. A. Teukolsky, W. T. Vetterling and B. P. Flannery, "Numerical Recipes in C, Second Edition", Cambridge University Press, Chapter 7, 1994.

[6] E. Cinlar, "Introduction to Stochastic Processes", Prentice-Hall, Chapter 5, 1975.

Fig. 6 reveals the waveform synthesis algorithm that uses constant RAM memory size. The current array, car, holds the isolated transition response spread over M clock bits

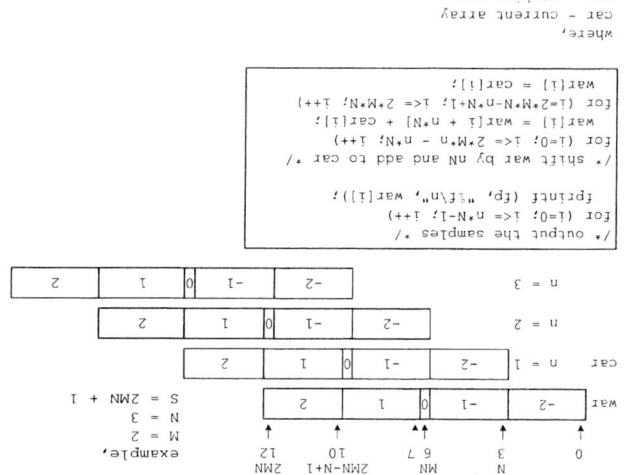

Fig. 6. Magnetic media waveform synthesis algorithm and pseudo-code.

where,
car - current array
war - working array
N - number of samples per symbol interval
M - number of symbol intervals per transition response on each side
S - size of car and war (2MN+1)
n - number of shifted symbol intervals (less or equal to 2M)

```
/* output the samples */
for (i=0; i<= n*N-1; i++)
  fprintf (fp, "%f\n", war[i]);

/* shift war by n*N and add to car */
for (i=0; i<= 2*M*N - n*N; i++)
  war[i] = war[i + n*N] + car[i];
for (i=2*M*N-n*N+1; i<= 2*M*N; i++)
  war[i] = car[i];
```

Fig. 5 depicts the method to determine the transition state. Our model uses two state variables $(s2,s1)$ to describe timing relationship of the current transition and possible transitions within the two previous bits. Since the system is causal, the media process activates at current time t ("now") whenever there is an event at the incoming write current data. Time stamps of the two preceding transitions are remembered as t_1 and t_2. Nominal clock period, T, is known, but not its instantaneous value, due to jitter/wander and WPC pre-compensation action. The transition state is determined by observing whether there was a transition within each of the two observation windows $(t - 2.5T, t - 1.5T)$ and $(t - 1.5T, t - 0.5T)$.

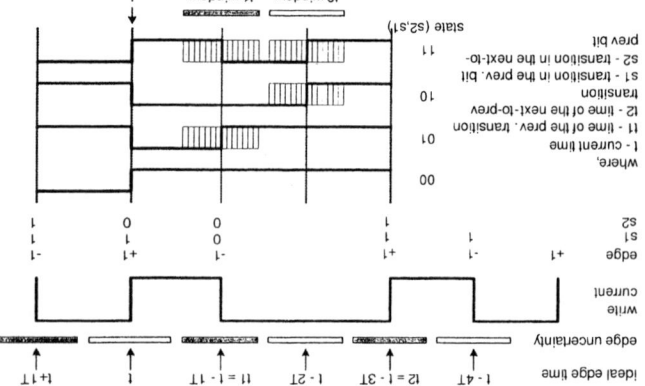

Fig. 5. Magnetic media model – state determination.

PID-CONTROLLED PLL FOR FAST FREQUENCY-HOPPED SYSTEMS

Hayri Uğur UYANIK, Nil TARIM

Istanbul Technical University, Electronics and Communication Engineering Department, 34469 Istanbul, TURKEY

uuyanik@itu.edu.tr, ntarim@ieee.org

Abstract— In this work, a novel aided-acquisition technique based on proportional-integral-derivative (PID) control of a phase-locked loop (PLL) is presented. This is achieved by inserting a control block into the PLL during acquisition where originally the output frequency/phase is controlled by a PI controller. This significantly reduces the settling time of the PLL which is important for certain applications such as WLANs where fast frequency-hopped spread-spectrum methods are used. Simulations show that the proposed structure reduces the setting time by 75%.

I. INTRODUCTION

Phase-locked loops (PLL) have many different applications especially in communications systems where they are used for carrier synchronization, clock recovery, frequency division/multiplication, demodulation and frequency synthesis. Depending on the application, the PLL often has different design considerations and specifications which in general constitute tradeoffs. In order to lighten the burden, aided-acquisition is one way to go, which is a technique based on changing the loop parameters during acquisition and restoring them afterwards [1, 2]. In applications such as WLANs where fast frequency-hopped spread-spectrum methods are used, the setting time of the PLL is of great importance. Therefore a method which improves the settling time but does not disturb the noise performance and power consumption would be very beneficial. A common way to achieve reduced setting times is to build PLLs with variable loop bandwidths [3–5]. Although this method does reduce the setting time, switching between bandwidths results with discontinuity in locking, with a possibility to lose lock.

In this work, an aided-acquisition technique based on the proportional-integral-derivative (PID) control of a PLL is proposed. The technique is applied by inserting a control block into the PLL during acquisition where originally the output frequency/phase is considered to be controlled by a PI controller. This addition of the control block changes the nature of regulation to a PID control and consequently reduces the setting time of the PLL. In order to avoid significant changes in noise performance and power consumption, the technique is applied during acquisition only. MATLAB simulations confirmed by Cadence simulations show that the proposed structure reduces the settling time by upto 75%.

II. PROPOSED PLL STRUCTURE

The traditional charge pump PLL, which excludes the control block shown inside the box with dashed lines, is depicted in Fig. 1. It comprises the phase/frequency detector (PFD), charge pump (CP), loop filter (LF), voltage-controlled oscillator (VCO) and divider (by N) blocks, where eventually frequency and phase acquisition is attained, thus $f_{out} = N \cdot f_{in}$ and $\Phi_{out} = N \cdot \Phi_{in}$. In this feedback system, the VCO output frequency/phase is controlled by the remaining blocks (except the divider) which constitute a PI controller. If, during acquisition, the control block in Fig. 1 is inserted into the loop, the PI control becomes PID type where the control current I_{ctrl} modifies the loop dynamics. However, there is no switching involved, and the injected control current gradually goes to zero as the phase error decreases. Since as the PLL locks, the control block smoothly rather than instantly stops injecting a control current, there is almost no danger of losing lock. Also, since the control block is disengaged after lock occurs, other parameters of the PLL remain unchanged if not improved.

In order to reap the advantages of the proposed structure, the control block should satisfy the following transfer function:

$$H_C(s) = \frac{I_{ctrl}(s)}{\phi_e(s)} = sK_a + K_b \tag{1}$$

where Φ_e denotes the phase error at its input, and K_a, K_b are coefficients characterizing the block.

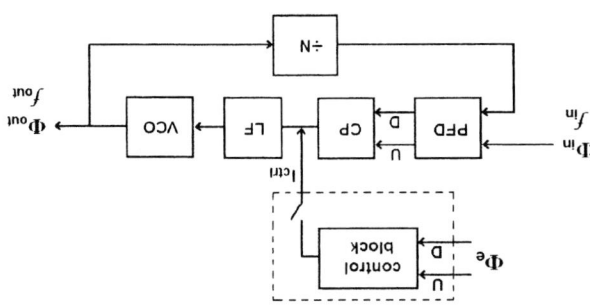

Figure 1. Proposed PID-controlled PLL structure

structure in Fig. 1 is

Provided that the loop filter consists of a series RC combination, the forward gain of the complete PLL

$$G(s) = (K_{PFD-CP} + H_C)\left(R + \frac{1}{sC}\right)\frac{K_{VCO}}{s} \quad (2)$$

where K_{PFD-CP} and K_{VCO} are the combined PFD-CP and VCO gains, respectively. Since all blocks of the forward path except the VCO are considered as a PID controller, the control parameters of the system become

$$K_P = (K_{CP} + K_b)R + \frac{K_a}{C} \quad (3a)$$

$$K_I = \frac{K_{CP} + K_b}{C} \quad (3b)$$

$$K_D = K_a R \quad (3c)$$

Equations (2) and (3) yield the corresponding closed-loop transfer function of the PLL as follows:

$$H(s) = N\frac{As^2 + 2\zeta\omega_n s + \omega_n^2}{s^2 + 2\zeta\omega_n s + \omega_n^2} \quad (4)$$

where the natural frequency ω_n, damping factor ζ, and the coefficient A can be expressed as

$$\omega_n = \sqrt{\frac{K_I K_{VCO}}{N + K_D K_{VCO}}} \quad (5a)$$

$$\zeta = \frac{1}{2}K_P\sqrt{\frac{K_{VCO}}{K_I(N + K_D K_{VCO})}} \quad (5b)$$

$$A = \frac{K_D K_{VCO}}{N + K_D K_{VCO}} \quad (5c)$$

A brief examination of (3), (4) and (5) yields that introducing the control block adds a derivative factor K_D into the controlling body as well as providing flexibility to the loop dynamics through K_P and K_I.

However, some practical considerations should be taken into account when designing the circuitry which satisfies the transfer function given in (1) such as: a) Since the phase error at the PFD output has discrete values, it should be filtered, b) A lossy derivative block should be used in order to ensure stability as well as maintaining smooth waveforms, c) The corresponding noise bandwidth for (4) becomes infinite which is intolerable for most PLL applications. Therefore, the injected current I_{ctrl} must be disabled after lock occurs.

As a result, one way of implementing the control block is given in Fig. 2. The working principle and how each sub-block contributes to the gradual diminishing of the control current can be explained as follows:

- The first order filter takes the voltage difference between the U and D signals and filters the difference so that the amount of error is achieved.
- The PD controller gets the amount of error as an input. It gives the sum of the first derivative of the input and the input itself, each multiplied with coefficients.
- The V/I converter gets the output voltage of the PD controller and generates the control current.
- The comparator compares the output of the PD controller with an interval and controls the switch accordingly.
- The switch simply transfers the control current to the control block output if the error is high or behaves like an open circuit if the error is low, thus making the PLL more reactive to changes without contributing additional noise.

Obviously limiting factors such as noise, stability, power consumption should be taken into account when determining the PID control parameters given in (3) and coefficients K_a, K_b. It goes without saying that the requirements which allow a continuous-time approximation of the PLL should be met [2, 6].

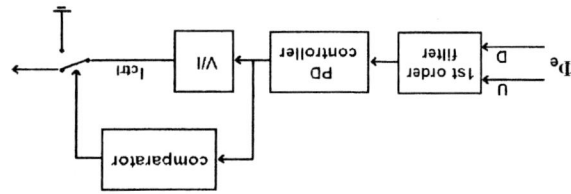

Figure 2. Implementation of the control block in the proposed PLL structure

III. TRANSIENT PERFORMANCE OF THE PLL

As mentioned before, our main interest lies on the settling time of the PLL. Straightforward analysis yields the transient response of the VCO voltage for a frequency step input by taking the inverse Laplace transform of the expression below:

$$V_{VCO}(s) = \Phi_{in}(s)H(s)\frac{K_{VCO}}{s} \quad (6a)$$

$$V_{VCO}(t) = \frac{NA\omega_{in}}{K_{VCO}}\left\{1-(1-A)e^{-\zeta\omega_n t}\left[\cosh(\omega_d t) - \frac{\zeta\omega_n}{\omega_d}\sinh(\omega_d t)\right]\right\} \quad (6b)$$

where $\Delta\omega_{in}$ is the amplitude of the frequency step input and $\omega_n = \omega_n\sqrt{\zeta^2-1}$. Using (6b), provided that S is the settling window (S=0.01) means the output voltage settles to a value within 1% of its final value, the settling time for $0<\zeta<1$ can be formulated as

$$t_s = \frac{1}{\zeta\omega_n}\ln\frac{A-1}{S\sqrt{1-\zeta^2}} \qquad (7)$$

Obviously, the corresponding settling times can be found for $\zeta = 1$ and for $\zeta > 1$ in a similar manner. As anticipated, (7) shows that the settling time is improved by the introduction of parameter A which stems from the control block added.

Most PLL designs have unity damping factors for optimized transient responses. On the other hand, it is known that $\zeta = 0.5$ yields optimum noise bandwidth for constant ω_n. Moreover, increasing ω_n decreases both settling time and pull-in time, however it also increases the noise bandwidth. These factors should be kept in mind during the design process.

IV. SIMULATION RESULTS

In order to demonstrate the versatility of the structure, the technique was applied on a 2.4 GHz CMOS frequency synthesizer employing a PLL characterized by the following parameters [7]: $I_{Cp} = 10$ μA ($K_{PFD-CP} = I_{CP}/2\pi$), $R = 12.15$ KΩ, $C = 28.9$ pF, $K_{VCO} = 6$ GHz/V, $N = 64$, which yield $\omega_n = 5.69$ Mrad/s, $\zeta = 1$, $K_P = 1.93\times10^{-2}$ V/rad, $K_I = 5.51\times10^4$ V/s.

Considerations on noise, stability and power consumption as well as the continuous-time approximation impose certain limits on ω_n, A, K_n, K_b. As a result, parameters characterizing the PID-controlled PLL become as follows: $\omega_n = 7.82$ Mrad/s, $\zeta = 1.5$, $A = 0.74$ and $K_n = 400\times10^{-15}$ A·s/rad^2, $K_b = 10^{-5}$ A/rad, $K_P = 15.46\times10^{-2}$ V/rad, $K_I = 4\times10^5$ V/s, $K_D = 4.86\times10^{-9}$ V·s/rad^2.

Cadence simulations showing the shift in the VCO input voltage for a frequency step input of 5 MHz and settling window of S=0.01 is given in Fig. 3a and 3b where the effect of the control block is demonstrated compared to its void. The gradual diminishing of the control current as lock occurs can also be observed in Fig. 4. It is obvious that the proposed technique reduces the settling time significantly from 1.114 μs to 293 ns which corresponds to an improvement around 75%.

V. CONCLUSION

In this work, a novel method is proposed which reduces the settling time of a PLL, which is important for applications where fast frequency hopping is used. The technique is based on the PID control of a PLL by inserting a control block into a traditional loop during acquisition. Simulation results show that the method enables reducing the settling time upto 25% of its initial value.

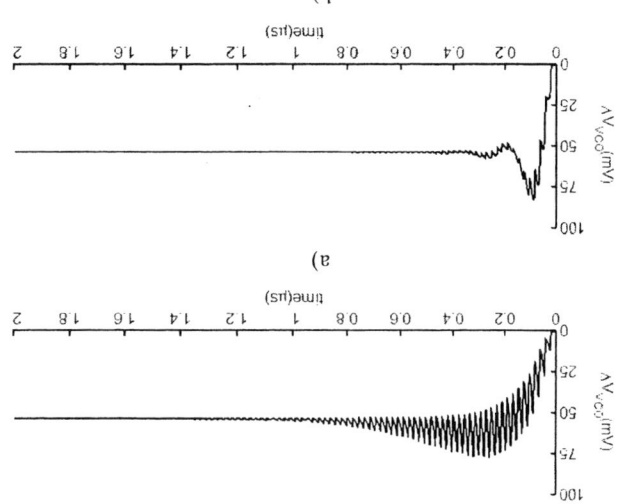

Figure 3. VCO input voltage deviation for a frequency step input of 5 MHz a) without the control block b) with the control block

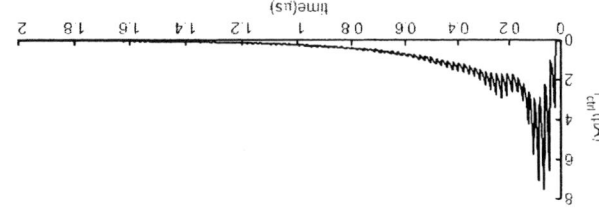

Figure 4. Variation of control current as the PLL locks

REFERENCES

[1] P. V. Brennan, "Technique allowing rapid frequency sweeping in aided-acquisition phase-locked loops", IEE Proceedings-G, 138, (2), pp. 205-209

[2] K. Shu, E. Sánchez-Sinencio, "CMOS PLL synthesizers: analysis and design", (Springer SBM, Inc, U.S.A, 2005)

[3] R. Dulk, "Digital fast acquisition method for phase-lock loops", Electronics Letters, 24, (17), pp. 1079-1080

[4] H. Sato, K. Kato, T. Sase, "A fast pull-in PLL IC using two-mode pull-in technique", Electronics and Communications in Japan, Part 2, 75, (3), pp. 41-50

[5] J. Lee, B. Kim, "A 250 MHz low jitter adaptive bandwidth PLL", Proceedings of IEEE Solid-State Circuits Conference, pp. 346-347

[6] F. M. Gardner, "Charge-pump phase-locked loops", IEEE Transactions on Communications, 28, (11), pp. 1849-1858

[7] B. Pak, "2.4 GHz CMOS PLL frequency synthesizer", (M.S. thesis, Istanbul Technical University, Institute of Science and Technology, 2002)

120

Optimum Design of Cascaded Digital Filters in Wideband Wireless Transmitters using Genetic Algorithms

Viral K. Parikh and Sankalp S. Modi, Poras T. Balsara

Department of Electrical Engineering, University of Texas at Dallas

Abstract—**In digitally intensive direct conversion transmitters, the baseband data is up-sampled to the RF rate. As the bandwidth of the baseband data increases, carefully designed cascaded digital filters are required in order to attenuate the wide replicas generated during the digital up-sampling process. Though design methodologies for single stage digital filters are very well established, near-optimum design of such multistage filters typically requires selection of several parameters by trial-and-error. This approach is time-consuming and does not assure the optimum solution (i.e. lowest area/power) under various performance constraints. In this paper, a genetic algorithm (GA) based generic automated search methodology for design of such cascaded filters is proposed. The proposed technique is demonstrated for digital WiMAX and WCDMA transmitters, for which near-optimal solutions appear to have been achieved in a relatively short time compared to the traditional manual design techniques.**

I. INTRODUCTION

A. Motivation

Recently proposed zero-IF digitally intensive transmitters achieve mixing and D/A operations at the RF rate using the digital circuitry [1], [2]. Such transmitters require digital up-sampling of the baseband data to the RF rate before the mixing operation. The Up-Sampling Ratio (USR) required in these transmitters are likely to be in the range of 16 to 32.

Figure 1 shows the spectrum of a GSM/EDGE standard compliant baseband signal, whose bandwidth is around 200 kHz. After up-sampling with an USR of 4, it can be seen that Zero-Order-Hold (ZOH) or linear interpolaters place notches at the integer multiples of the input sampling frequency and attenuate 400 KHz wide replicas. Figure 2 shows the spectrum of a 5 MHz wide baseband signal, which is typical for WLAN or WiMAX applications, after up-sampling with an USR 4. It is clear that the notches due to ZOH/linear interpolaters are not wide enough to suppress the wideband replicas. Our further experiments indicate that simple up-sampling and interpolation are insufficient for any broadband wireless standard, including WCDMA, WLAN and WiMAX, to attenuate the wideband replicas and thus additional filtering is essential.

Typically, area and power of such filtering are much lower, if it is implemented in the multiple stages rather than a single stage. Table I compares the area requirements for a single stage filter and a multistage filter to achieve attenuation of 90 dB at the first replica, when up-sampling a 5 MHz wide baseband signal with an USR of 32. As cascaded filters occupy 83% of less area, multistage designs are preferable over single stage designs for such stringent specifications. Cascaded filters are also used to achieve decimation operation with lower hardware complexity in receivers [3], [4].

Design methodologies for designing single stage digital filters are relatively very well established and automated. EDA tools such as a Matlab FDA tool [5] quickly provides the required order and the filter coefficients for the specified spectral response and the selected filter structure. However, design of cascaded filters is not straightforward. The designer needs to carefully select the appropriate values of the following parameters for each stage to minimize the total area before any of the single stage methodologies can be used: attenuation, output bit-length, coefficients bit-length, USR and filter structure. This selection is typically accomplished by trial-and-error or heuristics. These approaches are time-consuming and do not guarantee near-optimum solution. Moreover, the design cycle may need to be repeated each time for a different specification. An alternate approach is to utilize an automated search algorithm to obtain a near-optimum solution and reduce the design time.

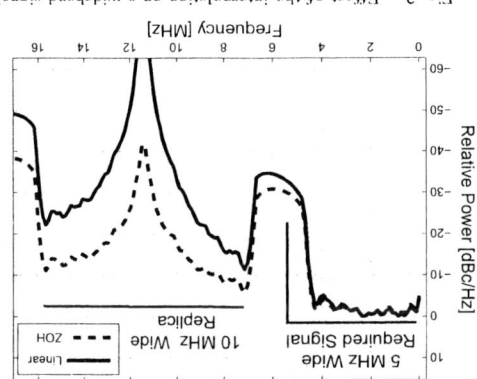

Fig. 1. Effect of the interpolation on a narrow-band signal

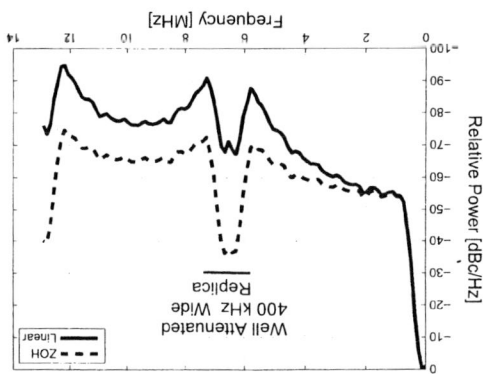

Fig. 2. Effect of the interpolation on a wideband signal

B. Generic Algorithms

The Genetic Algorithms (GAs) are population based blind, non-deterministic gradient-descent search algorithms, which have been successfully applied to a wide variety of the optimization problems [6]. Figure 3 shows a process of solution search in a simple GA, in which each solution is encoded in a fixed length of binary bits, termed as a 'genome'.

Initially genomes (design solutions) are generated randomly to form a pool of initial 'population' of the solutions. Each solution in the population is evaluated by a user-defined cost function (i.e. evaluation function) and is assigned a relative cost (i.e. fitness). In the selection process, a fixed number of least costly solutions are selected for 'reproduction'. The crossover process combines the good features of these selected solutions and generates a new set of solutions. The mutation process randomly flips some bits in the population to introduce diversity of the solutions in successive generations.

A GA gradually improves average quality of the solutions and the process is repeated until the termination criterion is met, which can be a fixed number of generations, the desired cost or convergence of the solutions. As a GA is considered quite efficient in exploring large search spaces [6], [7] and is unlikely to get trapped in a local minima, this algorithm is very useful for an automated blind search methodology to design the multistage filters. Additionally, a GA does not require the precise mathematical formulas of the objective function and is very generic, with only two problem specific aspects: 1) Encoding-decoding of a genome in a fixed bit-length, 2) Cost function to assign the relative cost to every solution.

II. THE PROPOSED DESIGN METHODOLOGY

A. Problem Formulation

The design of the n-stage cascaded digital filters is considered as an optimization problem with an objective of minimizing the implementation area. The performance constraints of the cascaded filter system is typically specified by a spectral mask and the acceptable quantization noise floor. Therefore, this problem can be viewed as the single objective (i.e. minimum area) with multiple constraints (i.e. spectral attenuation and quantization noise). The structures of all the filter stages are assumed to be fixed and identical (IIR Direct type-II). As the proposed design methodology is not implementation specific, any other filter structure can be used.

We consider the following four design parameters for each of the n different stages:

1) Attenuation (Att)
2) Output bit-length (O_{BW})
3) Coefficient bit-length (C_{BW})
4) USR (Sampling frequency)

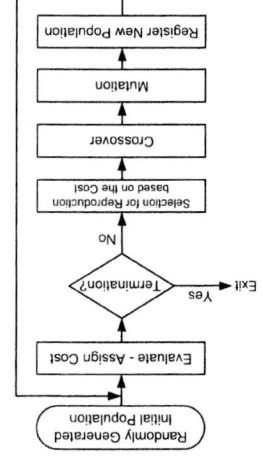

Fig. 3. Solution search in a GA

We further assume that once these parameters are chosen, each filter stage can be independently designed using a FDA tool and their combined performance can be evaluated in Matlab or other similar simulation tool [5]. Thus, the objective of the proposed design methodology is to select these $4n$ design parameters optimally.

B. Genome Representation

Each genome (design solution) is encoded into a fixed bit-length and it contains values of $4n$ design parameters. The number of bits allocated to each design parameter depends on the required range and the resolution (Δ) of the respective parameter. If δ is the difference between the maximum and the minimum range, bit-length (val) is calculated by eq. 1,

$$val = \lceil log_2(\delta/\Delta) + 1 \rceil \qquad (1)$$

For example, if the required range of the attenuation is from 40DB to 75dB, the resolution (step size) of 5dB, the bit-length required to represent this parameter is 3, as shown in Figure 4. The 3-bit in the genome can take eight possible values, which reflects one of the eight attenuation numbers {40,45,50,55,60,65,70,75} in the cost function.

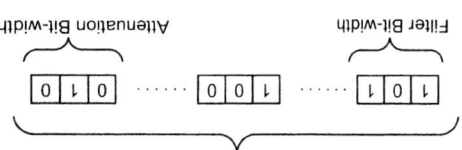

Fig. 4. An example of a genome bit stream

TABLE I
AREA COMPARISON FOR SINGLE VS. MULTISTAGE FILTERS

Stage	USR	Attern. at 1st Replica [dB]	Order/ Section	Bit Len.	Norm. Area
Single	32	90	12/6	15	1
Multi-1st	2	30	4/2	10	0.06
Multi-2nd	4	40	3/2	12	0.07
Multi-3rd	4	50	2/1	15	0.04
Total	32	-	-	-	0.17

C. Cost Function

A GA selects and reproduces increasingly better solutions based on the relative estimated cost of the present solutions. For every design solution D, a cost function decodes the genome into selected parameter values. With these selected parameters, the area requirements and its compliance against the noise specifications of the targeted standard are estimated.

1) Area Estimation: A GA requires only relative cost of the estimated area and does not need the accurate area calculation in the absolute unit (i.e. mm^2). Now, the area of a digital filter depends on the total number of registers (delay elements), adders and multipliers used in its implementation. A Matlab FDA tool provides order and coefficients for each stage based on the selected spectral attenuation and sampling frequency. Thus, the area cost of each stage is estimated based on:

1) Decoded output bit-length (O_{Bw})
2) Required number of coefficients for each stage provided by a Matlab FDA tool (no)
3) Decoded coefficient bit-length (C_{Bw})

The following formulas are used to estimate the total area (A_{Est}) for a given design solution D,

$$A_{Reg} = no \cdot O_{Bw} \cdot U_{Reg}$$
$$A_{Add} = no \cdot O_{Bw} \cdot U_{Add}$$
$$A_{Mult} = no \cdot O_{Bw} \cdot C_{Bw} \cdot U_{Add}$$
$$A_{Est} = A_{Reg} + A_{Add} + A_{Mult} \quad (2)$$

U_{Add} and U_{Reg} denote area of a single register and an adder, respectively.

2) Evaluation of Spectral Noise: Matlab simulations are used to verify the compliance of a given design solution, D, against various noise specifications, which is utilized as the design constraints. The actual baseband data, compliant to the targeted standard, is passed through the cascaded filter stages, which are designed according to the decoded parameters and the filter coefficients, for a given solution D. A Matlab function verifies the close-in noise performance against the transmitter spectral mask, defined in the targeted standard, and also evaluates the far-out noise compliance. The detailed examples are given as the case studies in Section III.

3) Constraint Handling in a GA: For an efficient constraint handling, the cost function with a penalty term described in [8] is adopted. Assume that for the targeted wireless standard, the tolerable limits of the close-in and the far-out noise are $L_{Close-in}$ and $L_{Far-out}$, respectively. Also, the simulated measured noise is $N_{Close-in}$ and $N_{Far-out}$ for the design D, corresponding to the decoded genome. Therefore, the constraints become $g1 = N_{Close-in} - L_{Close-in} < 0$ and $g2 = N_{Far-out} - L_{Far-out} < 0$. If the estimated area of the design D is A_{Est}, the overall cost function $Cost$ is given by:

$$Cost = \frac{A_{Est}}{A_{Max}}, \quad if \ g1 < 0 \ and \ g2 < 0$$
$$= 1 + \frac{p(g1) + p(g2)}{N_{Max}}, \quad otherwise \quad (3)$$

where, $\quad p(x) = x, \quad if \ x > 0$
$$= 0, \quad otherwise$$

Both the area and the noise constraints are normalized with respect to the maximum area (A_{Max}) and the maximum noise (N_{Max}), respectively. This cost function ensures that designs violating any of the constraints will be assigned higher cost compared to all the designs that do not violate the constraints. Also, the value of the penalty will be proportional to the amount of the constraint violations.

III. CASE STUDIES

As case studies, design of multistage digital filters for a WiMAX and a WCDMA transmitter is presented. Figure 5 shows two 3-stage cascaded digital filters, used in such digital transmitters, to increase the sampling rate of the baseband signal while attenuating the associated replicas. The goal is to design the multistage filter optimally by choosing 12 design parameters (4 × 3) to minimize the overall implementation area. We performed experiments for each of the transmitters by using a deterministic tournament search based genetic algorithm with population size, crossover rate and mutation rate of 60, 0.7 and 0.1, respectively.

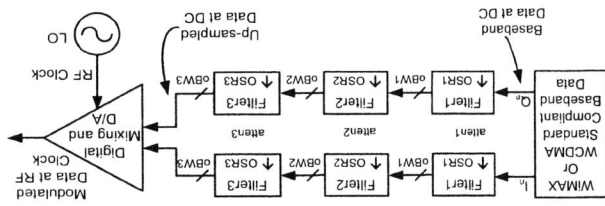

Fig. 5. 3-stage digital filters used in the digital transmitters.

A. Selection and Mapping of Design Parameters

Table II presents the selected design parameters, their ranges, suitable step sizes (resolution-∇) and required bit-length for a genome representation, as per eq. 1. The last column of Table II gives mapping of bit-length values (val) to the actual design parameter value, which is given by equation $val \cdot \nabla + min(range)$. As each stage of the cascaded filters requires a 9-bit genome representation, total length of a genome for the 3-stage filters is 27. The required USR is 32.

TABLE II
RANGE AND MAPPING OF DESIGN PARAMETERS

Design Parameters	Range {g}	Step (∇)	Bit-Length (∆)	Mapping (val)
Attenuation [dB]	30-58 {28}	4	3	$val \cdot \nabla + 30$
Output bit-length	8-15 {7}	1	3	$val \cdot \nabla + 8$
Coeff. bit-length	10-16 {6}	2	2	$val \cdot \nabla + 10$
USR	2-4 {2}	2	1	$val \cdot \nabla + 2$

REFERENCES

[1] R. B. Staszewski, J. Wallberg, S. Rezeq, C.-M. Hung, O. Eliezer, and et. al., "All-digital PLL and transmitter for mobile phones," *IEEE Journal of Solid-State Circuits*, vol. 40, no. 12, pp. 2469–2480, 2005.

[2] P. Eloranta and P. Seppinen, "A WCDMA transmitter in 0.13um CMOS using direct-digital RF modulator," 2007, pp. 240–241.

[3] R. R. Shively, "On multistage finite impulse response (fir) filters with decimation," *IEEE Transactions on Acoustics, Speech, and Signal Processing*, vol. ASSP-23, no. 4, pp. 353 – 357, 1975.

[4] Y.-C. Lim, R. Yang, and B. Liu, "Design of cascaded fir filters," vol. 2, Atlanta, GA, USA, 1996, pp. 181 – 184.

[5] http://www.mathworks.com/products/filterdesign/. The Mathworks.

[6] L. e. Davis, *Handbook of Genetic Algorithms*. Van Nostrand Reinhold, New York, NY, 1991.

[7] P. Singh, "Comparative study of genetic algorithm and simulated annealing for optimal tolerance design formulated with discrete and continuous variables," *Proceedings of the Institution of Mechanical Engineers, Part B*, vol. 219, no. 10, pp. 735-760, October 2005.

[8] K. Deb, "Efficient constraint handling method for genetic algorithms," *Computer Methods in Applied Mechanics and Engineering*, vol. 186, no. 2, pp. 311-338, June 2000.

[9] *IEEE 802.16 standard; Air Interface For Fixed Broadband Wireless Access Systems*, October 2004.

[10] *3G TS 34.21 v3.0.1 3rd Generation Partnership Project*, 2003.

IV. CONCLUSION

We presented an automated design methodology for the cascaded digital filters, used to attenuate the wideband replicas during digital up-sampling. In the presented case studies, the proposed GA based methodology provided design solutions with minimum area while satisfying the design constraints and avoiding any over-design, offering a rapid design technique to system designers. Though this paper only examines cascaded filters design for a specific problem in wireless transmitters, the proposed design methodology is quite generic and can be easily adopted for the other applications of cascaded filters, as presented in [3], [4]. As the final outcomes are not very sensitive to the parameter settings of a GA, designers do not require special knowledge of the algorithm for its application. A GUI interface for the ease of a designer is planned for the future work.

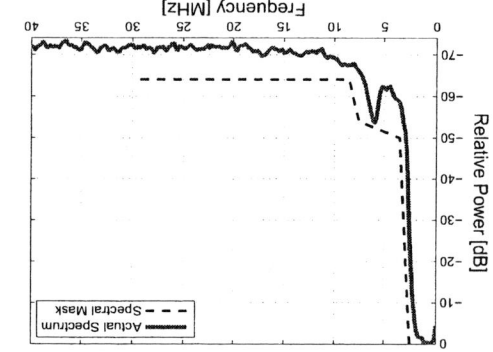

Fig. 7. Resultant spectrum and spectral mask for WCDMA

$$Noise@(5.45, 9.75, 14.75, Far)MHz = -(25, 32, 50, 60)dB \quad (4)$$

$$Noise@(2.5, 3.5, 7.5, Far_{out})MHz = -(50, 54, 64, 70)dB \quad (5)$$

B. Noise Constraints

The transmitter spectral masks, defined in WiMAX standard [9] and WCDMA standard [10] and repeated in eq. 4 and 5, respectively, are used as performance constraints. These constraints were evaluated with the corresponding standard compliant baseband data, i.e. 5 MHz wide for WiMAX and around 2 MHz wide for WCDMA, using Matlab simulations.

C. Results

Table III presents the best solutions found by the proposed design methodology for WiMAX and WCDMA, respectively. Each experiment was repeated with different sets for noise specification changes. The solutions, presented in the second column, were verified for the noise specification compliance, given in the first column. The spectrums of the best solutions, presented in the rows 1 and 4, are plotted in Figure 6 and 7, respectively. It can be observed that the solutions found by the proposed GA based design methodology satisfy the constraints and yet do not leave any unnecessary margins between the noise specifications and actual responses. The execution time for each GA run was between 110 to 170 minutes.

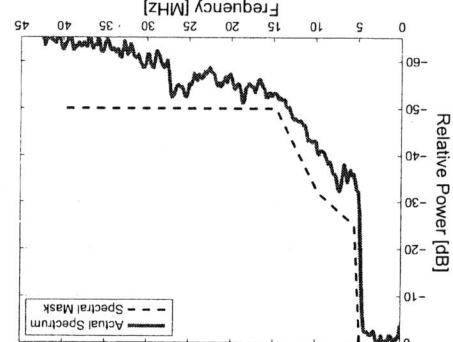

Fig. 6. Resultant spectrum and spectral mask for WiMAX

TABLE III
SIMULATION RESULTS FOR WIMAX AND WCDMA

@(Frequencies) [MHz] = (Noise Limits) [dB]	$Art_{1,2,3}$,$O_{Bw}1,2,3$, C_{Bw}=1,2,3,USR1,2,3	Normalized Area
WiMAX		
@(5.45,9.75,14.75,Far) =−(25,-32,-50,-50)	30,30,34,8,10,11. 12,12,12,4,2,4	0.1903
@(5.45,9.75,14.75,Far) =−(30,-37,-55,-60)	30,42,50,8,9,13. 12,12,12,4,2,4	0.19377
@(5.45,9.75,14.75,Far) =−(35,-42,-65,-65)	42,58,50,8,12,14. 12,12,12,4,2,4	0.25375
WCDMA		
@(2.5,3.5,7.5,8.5,Far) =−(50,-54,-64,-64)	34,38,46,8,8,8. 12,12,12,4,2,4	0.1385
@(2.5,3.5,7.5,8.5,Far) =−(60,-64,-74,-80)	34,46,50,8,9,11. 12,12,12,4,2,4	0.1585

A 400MHz-2.4GHz Radiation-Tolerant Self-Biased Phase-Locked Loop

Peiqing Zhu[1], Ping Gui[1], Wickham Chen[1], Annie C. Xiang[2], Datao Gong[2], Tiankuan Liu[2], Yanli Fan[3], Huanzhang Huang[3], Mark Morgan[3]

[1]Department of Electrical Engineering
[2]Department of Physics
Southern Methodist University

[3]Texas Instruments
Dallas, Texas

Abstract—This paper describes the design of a self-biased Phase-Locked Loop for radiation-tolerant applications. A novel single-to-differential converter circuit that eliminates the mismatches on the output of a phase-frequency-detector is proposed to minimize reference spurs. Design considerations for radiation-tolerant design are also described. Fabricated in a 0.25 μm CMOS Silicon-on-Sapphire process, the PLL achieves an operating frequency range from 400MHz to 2.4GHz. The RMS jitter of the PLL is 3.9ps at 2.4 GHz.

Index Terms -- Phase-Locked Loop, Self-biased, Radiation Tolerant, jitter

I. INTRODUCTION

Applications such as aerospace, medical irradiation, nuclear instruments and high-energy physics experiments require circuits and systems to sustain various level of radiation. Radiation in general affects circuits in two major ways: the first effect, known as the total ionizing dose (TID), represents the cumulative effects of many particles hitting a device throughout the course of its mission life, gradually shifting the transistor threshold voltage, degrading the transistor matching, and the device performance until it ultimately fails. The second effect, referred to as single-event effects (SEE) involves high energy particles that penetrate deep into materials and components, leaving a temporary trail of free charge carriers. If these particles hit vulnerable spots in the circuit, they can produce adverse effects such as introducing bit errors in digital circuitry and perturbing analog circuitry [1].

A Phase-Locked Loop (PLL) is one of the most sensitive components for analog and mixed-signal circuits working in radiation environment. A typical PLL is shown in Fig. 1. It consists of a phase frequency detector (PFD), a charge pump, a loop filter, a voltage-controlled oscillator (VCO), and a frequency divider. The PFD compares the phase of the reference clock to that of the divider clock and generates *up* (U) and *down* (D) pulses based on the phase error. The U and D pulses turn on the current sources/sinks in the charge pump

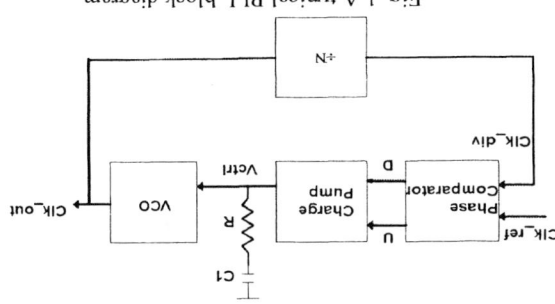

Fig. 1 A typical PLL block diagram.

to charge/discharge a capacitor, resulting in the control voltage Vctrl. Vctrl adjusts the VCO frequency until the phase error between the reference and divided clock is zero. When the PLL is in lock, the frequency of the output clock is N times that of the reference clock.

Typically, the SEE affects the digital circuitry in the PLL, such as PFD and divider. Whereas, the TID mostly affects analog circuitry including the VCO. Traditionally radiation-tolerant design relies on specialized fabrication processes which could be many orders of magnitude more expensive than commercial processes. Recently, the Radiation-Hard-by-Design (RHBD) approach has gained more popularity because it allows radiation-tolerant circuits to be built using commercially available low-cost CMOS processes [2] [3].

To mitigate the SEE in our PLL design, we choose a commercial 0.25μm Silicon-on-Sapphire (SOS) CMOS technology for its improved SEE immunity compared to the bulk CMOS technologies. In addition, we increased the transistor size in SEE sensitive circuits such as the divider and PFD so that they are less likely to be affected by SEE hits.

We explore novel circuit architectures and design techniques to mitigate the TID effects. Since the TID introduces variations on the transistor parameters, it is important to design the circuit in such a way that its operation is insensitive to those variations. For our PLL design, we choose a self-biased architecture which has an operating point independent of the process technology and environmental variations hence rendering a more robust design. The self-biased architecture also gives a constant ratio of the bandwidth to reference frequency and a constant damping factor. These ensure the stability of a PLL for a large range of input frequencies while at the same time keeping the phase noise low across the entire reference frequency range.

In addition, we propose a special singled-ended-to-differential converter circuitry to minimize the phase noise caused by the mismatches in the PFD outputs. Fabricated in a 0.25 μm Silicon-on-Sapphire CMOS process, the PLL has an operating frequency range of 400MHz-2.4GHz. The RMS jitter of the PLL output is measured to be 3.9ps at 2.4GHz.

II. SILICON-ON-SAPPHIRE TECHNOLOGY

We chose Peregrine Semiconductor's Silicon-on-Sapphire (SOS) UltraCMOS® technology for its good SEE immunity. As shown in Fig. 2 [4], this SOS technology has a very thin layer of silicon (about 100nm) deposited on top of the sapphire substrate (about 200um). This thin layer is silicon where the transistor devices are made, isolated by LOCOS. Compared to bulk CMOS technologies, the SOS has a lower SEE sensitive volume. In addition, the SOS technology has lower power dissipation because of its reduced junction capacitance. Finally, the high resistivity of Sapphire substrate (10^{16} Ω-cm) helps reducing substrate crosstalk [4].

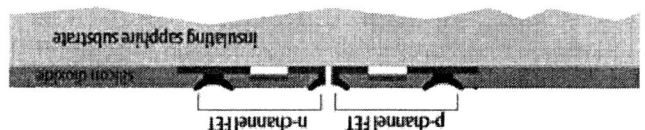

Fig. 2 SOS CMOS Technology [2].

We performed an SEE and a TID test on a test chip manufactured using this technology. Results show that there is no SEE-introduced bit error in the test structures under a fluence of 1.8×10^{12} proton/cm^2 [5]. This confirms the good SEE immunity of this technology. On the other hand, we observed that the PMOS transistors experience an increase in threshold voltage $|V_{tp}|$ during and after irradiation, whereas the NMOS transistors initially experience an increase in V_{tn} at lower total dose and a decrease in V_{tn} at higher total dose [5]. In addition, PMOS transistors see leakage current of 1uA right after irradiation whereas the leakage current for NMOS transistors after irradiation is negligible. A radiation-tolerant design must take the TID effects into considerations.

III. DESIGN OF THE PLL

A. Phase Frequency Detector (PFD)

We use a tri-state PFD in our PLL that detects both the frequency and the phase difference. When the PLL is in lock, the width of the UP/DN pulses is made to be about 300ps to minimize static phase error while at the same time eliminating the dead-zone problem.

B. Single-to-differential (S2D) converter

The charge pump utilizes UP/DN and their complementary signals to switch the current source and current sink. To minimize the static phase noise, it is important to minimize the mismatches between UP/DN and their complementary signals. A conventional design uses a transmission gate in one path and an inverter in the complementary path to match the delays between the two. In reality, it is difficult to perfectly match these two delays due to process and transistor parameters variations. We proposed a novel circuit to match the delays.

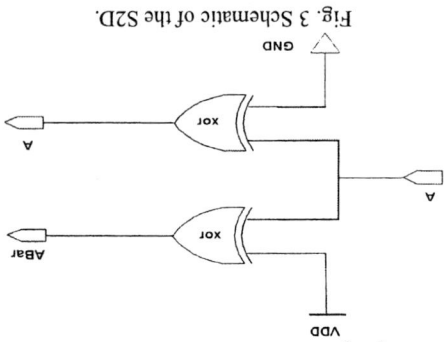

Fig. 3 Schematic of the S2D.

Figure 3 shows the schematic of the proposed circuit. It is made of two XOR gates. The single-ended signal (represented by A) is fed to both XOR gates. One of the XOR gate has Vdd as an input whereas the other has Gnd as an input. The XOR gate with a VDD as an input generates the complementary signal ABar while the other XOR gate generates the buffered version of A. Since the two XOR gates are identical, they introduce exactly the same delays in generating ABar and A, thus minimizing the mismatch in the charge pump operations.

C. VCO design

At the core of a self-biased PLL is a low-noise VCO. To minimize its sensitivity to power supply noise, temperature and radiation-induced transistor parameters variations, we designed the VCO with symmetrical load and a replica-feedback biasing [6].

The VCO consists of eight differential delay stages with symmetrical loads. Figure 4 shows the schematics of the delay cell [7]. Since the SOS technology has low substrate noise, we focus on minimizing the power supply noise in our design by choosing a source-coupled NMOS differential pair with PMOS symmetrical loads.

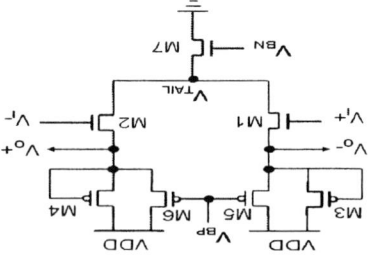

Fig. 4 Differential delay cell with symmetrical load.

The symmetrical load is composed of two equally-sized transistors in parallel, one in diode connection and the other controlled by V_{BP}. When the voltage swing across the load is controlled by V_{BP}, the I-V curve of the symmetrical load is approximately linear, giving rise to a high dynamic power supply noise rejection [7]. Figure 5 (a) shows the symmetrical load and (b) shows the simulated I-V curve.

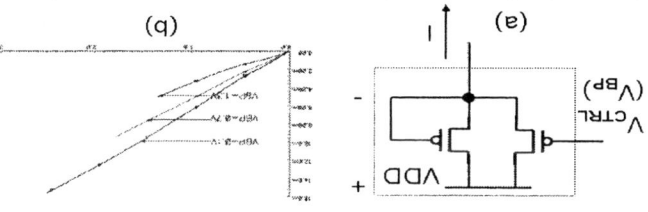

Fig. 5 (a) The symmetrical load; (b) The I-V curve of the symmetrical load [5].

To generate the correct voltage swing across the symmetrical load, the differential delay cell is biased using a replica-biasing circuitry as shown in Fig. 6 [7].

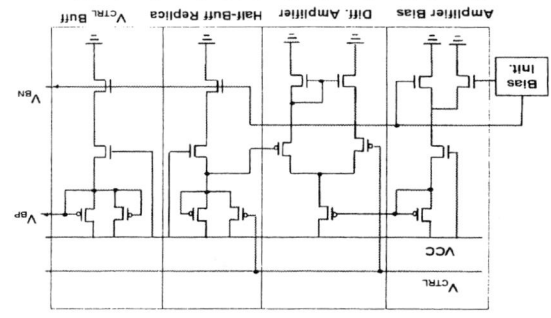

Fig. 6. Replica-feedback biasing circuit [5].

The self-biasing generator also dynamically adjusts the V_{BN} to provide a constant signal swing V_{BP} across the load in the presence of power supply noise and transistor parameters variations.

Simulation indicates that the VCO has a phase noise of -96dBc/Hz at 1MHz, as shown in Fig. 7.

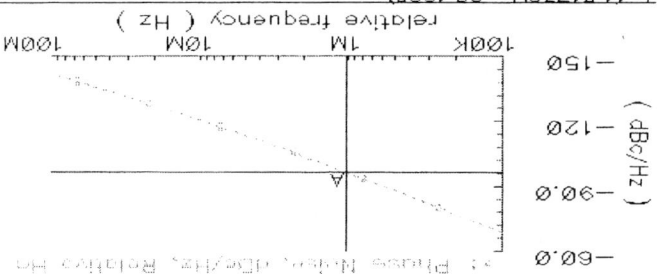

Fig. 7 VCO Phase Noise.

D. Charge Pump

The self-biased structure is also employed in the charge pump to produce a zero-offset static phase error [7]. The charge pump consists of two source-coupled pairs each with symmetrical loads and a current source. As shown in Figure 8, when both UP and DN signals are enabled, the current mirror structure (made of the transistors M1, M4, load L1 and L2 and the tail current source M5 and M6) ensures that the source and sink current are exactly the same. The symmetrical load and the V_{BN}-controlled current source in the left source-coupled pair acts like a half-replica, producing at the mirror node the same voltage as V_{BP} and making the current though the charge pump proportional to that in the VCO delay cell. This feature is utilized to generate a fixed damping factor ζ and a fixed bandwidth to operating frequency ratio, as will be explained in the next section.

Leakage current and mismatch between the current sources will introduce jitter and spurs on the PLL output. Since the leakage current for PMOS in the technology is on the order of 1uA after irradiation whereas that for NMOS is negligible, we increase the current through the charge pump to minimize the this mismatch. The current used in our charge pump is 3mA.

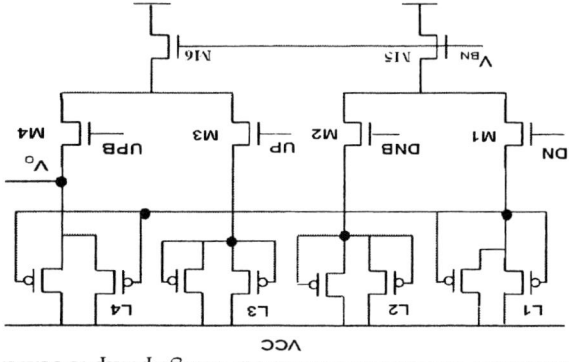

Fig. 8 The schematic of the charge-pump [5].

E. PLL architecture

For a second-order PLL, the damping factor ζ, the natural frequency ω_N and the 3-dB bandwidth are given by the following formulas [8][9][10].

$$\zeta = \frac{1}{2}\sqrt{\frac{1}{N} \cdot I_{CH} \cdot K_v \cdot R^2 \cdot C_1} \qquad (1)$$

$$\omega_N = \frac{2\zeta}{R \cdot C_1} \qquad (2)$$

$$\omega_{3dB} = \omega_N \left(1+2\zeta^2 + \sqrt{2+4\zeta^2+4\zeta^4}\right)^{\frac{1}{2}} \qquad (3)$$

It has been proven that by setting I_{CH} proportional to the VCO delay cell current I_v, and setting R reversely proportional to the square root of I_v, we can obtain a fixed constant ζ and $\omega_{3dB}/\omega_{ref}$ independent of the operating frequency, given by the following equations [7].

$$\zeta = \frac{y}{4} \cdot \sqrt[4]{\frac{x}{N}} \cdot \sqrt{\frac{C_1}{C_B}} \qquad (4)$$

$$\frac{\omega_N}{\omega_{REF}} = \frac{\sqrt{x \cdot N}}{2\pi}\sqrt{\frac{C_B}{C_1}} \qquad (5)$$

In the above equations, C_B is defined as $2nC_{EFF}$, with C_{EFF} being the effective output capacitance of the delay cell. x in equation (4) and (5) represents the ratio of the I_{CH} / I_v and y represents the ratio of loop filter resistance to the resistance of symmetrical load. To make y a constant, the loop filter R is produced as $1/g_m$ of the symmetrical load in the bias generator together with a second charge pump.

Figure 9 gives the overall structure of the self-biased PLL. In addition to two charge pumps, there are two bias-generators: one has a start-up circuit to ensure that the control voltage is always within a valid range, and the other one provides the actual V_{Bp} to the VCO.

In our design, damping factor ξ is set to be 1 and $\omega_{3dB}/\omega_{ref}$ is about 1/20. Simulations are performed to make sure the PLL has enough phase margin (>45°).

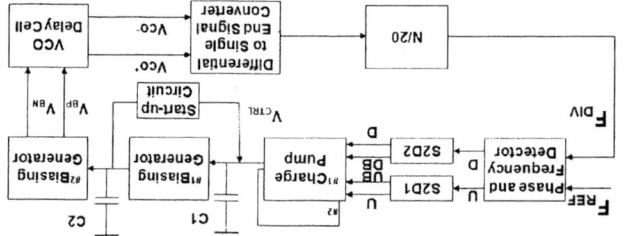

Fig. 9 Block diagram of the self-biased PLL.

F. Chip micrograph

Figure 10 shows the micrograph of the chip. The chip area is about 330 um².

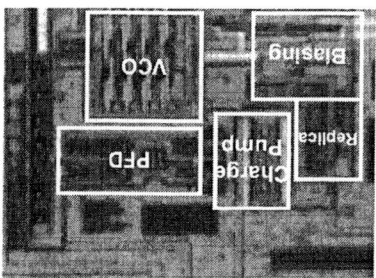

Fig. 10. Chip micrograph.

IV. MEASUREMENT RESULTS

The PLL is tested to be fully functional. Figure 11 shows the simulated and measured plots of PLL output frequency versus the control voltage. The frequency range is measured to be 400MHz to 2.4GHz.

The PLL jitter is measured at the PLL output. Figure 12 shows the histogram of the PLL clock output at 2.4GHz through a buffer circuit. The RMS jitter is measured to be 3.9ps whereas the peak-peak jitter is measured to be 48ps.

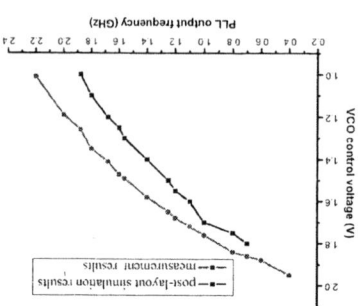

Fig. 11 Simulated and measured Freq. vs. V_BP

We plan to carry out the radiation tests on the PLL in the near future and the post-radiation measurement results will be presented at the workshop.

V. CONCLUSION

We designed a self-biased PLL for radiation-tolerant applications which is insensitive to environmental and radiation-induced transistor parameter variations and has a wide operating frequency range. A new single-to-differential converter circuit is used to minimize reference spurs. Our measurement results show that the PLL can achieve a 400MHz to 2.4GHz frequency operating range using a commercial 0.25µm SOS CMOS process. The RMS jitter is measured to be 3.9ps at 2.4 GHz.

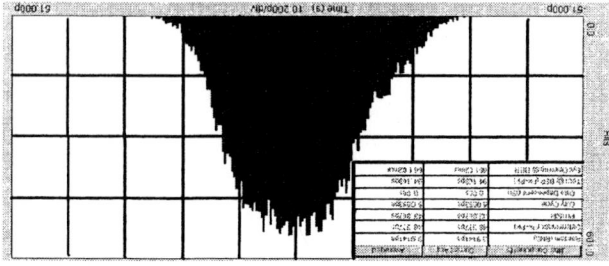

Fig. 12 Measured jitter histogram at a 2.4GHz clock output.

ACKNOWLEDGMENT

The authors thank NSF/ATLAS program for funding of the project and Peregrine Semiconductor Corp. for their support of this work.

REFERENCES

[1] C. Claeys and E. Simoen, *Radiation Effects in Advanced Semiconductor Materials and Devices*, Berlin Heidelberg, ISBN 3-540-43393-7, Springer-Verlag, 2002.

[2] D. R. Alexander, "Design issues for radiation tolerant microcircuits in space," IEEE Short Course presented at the 1996 NSREC Conference, Indian Wells, CA, July 15-19, 1996, V-1.

[3] G. Anelli, M. Delmastro et al., "Radiation Tolerant VLSI Circuits in Standard Deep Submicron CMOS Technologies for the LHC Experiments: Practical Design Aspect", IEEE Transactions on Nuclear Science, vol. 46, no. 6, December 1999, pp. 1690-1696.

[4] Peregrine Semiconductor Corp., "UltraCMOS Process Technology",http://www.psemi.com/content/foundry/foundry_process_tech.html.

[5] T. Liu, W. Chen, P. Gui, J. Yang, J. Zhang, P. Zhu, A. C. Xiang, J. Ye, R. Stroynowski, "Total Ionization Dose Effect and Single-Event Effect Studies of a 0.25 µm Silicon-On-Sapphire CMOS Technology", *Proceedings of The 13th NASA Symposium on VLSI Design*, June 2007.

[6] J. Maneatis and M. Horowitz, "Precise delay generations using coupled oscillators," *IEEE J. Solid-State Circuits*, vol. 28, no. 12, pp. 1273-1282, Dec. 1993.

[7] J. Maneatis, "Low-jitter process-independent DLL and PLL based on self-bias techniques," *IEEE J. Solid-State Circuits*, vol. 31, no. 11, pp. 1723-1732, Vol. 31, No. 11, November 1996.

[8] Floyd M. Gardner, *Phase Lock Techniques*, Wiley-Interscience, 1979.

[9] Thomas Lee, The Design of CMOS Radio-Frequency Integrated Circuits, Cambridge.

[10] Garth Nash, *Phase-Locked Loops Design Fundamentals*, Freescale Semiconductor Application Note, Document Number: AN535, Rev. 1.0, 02/2006.

A Speed- and Accuracy-Enhanced On-Chip Current Sensor with Local Shunt Feedback for Current-Mode Switching DC-DC Converters

Mengmeng Du and Hoi Lee
Department of Electrical Engineering
The University of Texas at Dallas
Tel: 1-972-883-4841   Fax: 1-972-883-2710   Email: hoilee@utdallas.edu

Abstract—A high-speed high-accuracy on-chip current sensor for current-mode switching dc-dc converters is presented in this paper. By employing local shunt feedback, the non-dominant pole in the current sensor is pushed to high frequency. Both speed and accuracy of the current sensor can then be improved simultaneously by consuming low quiescent current. The proposed current sensor for the buck converter is designed using a standard 0.35μm CMOS process. Results show that the proposed sensor achieves at least 95% sensing accuracy and <50ns settling time by consuming the quiescent current of 35μA. The sensor allows the current-mode buck converter to operate at switching frequencies up to 2MHz with a duty cycle down to 0.2.

I. INTRODUCTION

Switching-mode dc-dc converters are essential in all battery-operated portable electronic devices for converting a time-varying input battery voltage to a constant output voltage for supplying different internal systems. This is due to the reason that switching-mode power converters can perform highly efficient dc-dc conversion to maximize the battery run time. Generally, switching-mode power converters require the use of both off-chip inductor and output capacitor for delivering hundred milliampere output current in portable applications. It is critical to minimize the required inductance and capacitance in order to decrease the size and weight of the portable devices through miniaturization of the power modules. As a result, operating switching power converters at high switching frequencies is needed for decreasing the required inductance and capacitance. On the other hand, current-mode control has advantages of automatic over-current protection, better closed-loop stability and faster dynamic response over voltage-mode control. Although current-mode control is much better, a high-speed high-accuracy current-sensing circuit is required to sense the inductor current and allow current-mode switching-mode power converters operating properly at high switching frequencies. It is challenging to achieve both high speed and high accuracy simultaneously in the on-chip current sensor.

Different on-chip current sensors have been reported to realize the current-mode switching-mode power converters [1]-[7]. However, reported on-chip current sensors cannot simultaneously achieve over 90% sensing accuracy and allow the switching-mode power converters operating beyond 1MHz by consuming low quiescent current. In particular, the accuracy of the current sensor depends on the voltage mirror realized by bipolar transistors and hence the current sensor is implemented in BiCMOS process [1]. When the accuracy of the current sensor is enhanced by using a two-stage internal amplifier, the current-sensor speed is limited by the unity-gain frequency of the two-stage amplifier to operate beyond 1MHz under low quiescent condition [2]-[5]. With an offset current cancellation technique in [6], the high accuracy of the current sensor can be realized by using a single-stage amplifier. This technique reduces the required quiescent current to achieve high-speed operation as compared to that in [2]-[5]. However, the current sensor in [6] suffers from the tradeoff between the speed and the accuracy. The bias current of the current sensor should be increased to improve the speed, which decreases the loop-gain magnitude of the current sensor and thus lead to the degradation in the sensing accuracy. Besides, the lossless self-learning current-sensing technique requires complicated circuit implementation sense the inductor current, resulting in large dissipation in quiescent current and large chip area [7].

In order to address above issues, an on-chip current sensor with local shunt feedback is proposed in this paper. The local shunt feedback enables both sensing accuracy and speed to be optimized simultaneously. The local shunt feedback can push the non-dominant pole in the current sensor to a much higher frequency, thereby improving both the unity-gain frequency and phase margin of the current-sensor loop-gain transfer function. The speed of the current sensor can thus be enhanced under low quiescent-current condition. Good sensing accuracy can also be achieved, as the loop-gain magnitude of the current sensor is not affected by the local shunt feedback. Additionally, the local shunt feedback can be realized by two small transistors, which indicates that the proposed current sensor is a compact design. Table I provides the performance

TABLE I
PERFORMANCE COMPARISONS WITH PREVIOUSLY REPORTED CURRENT SENSORS

	[1]	[3]	[5]	[6]	[7]	This Work
Technology	BiCMOS	0.6μm CMOS	0.6μm CMOS	0.35μm CMOS	0.35μm CMOS	0.35μm CMOS
Year	1998	2004	2005	2004	2007	2007
Maximum Frequency (MHz)	1	1	0.5	0.5	0.78	2
Accuracy	N.A.	96%	94.4%	N.A.	91.5%	95%
Quiescent Current (μA)	N.A.	N.A.	N.A.	200	N.A.	35

comparisons of different current sensors. The proposed current sensor achieves the fastest speed with small quiescent current and good sensing accuracy as compared to other reported current sensors. This paper is organized as follows. Section II discusses the operational principle and local shunt feedback of the proposed current sensor. Implementation results and conclusions are given in Sections III and IV, respectively.

II. PROPOSED CURRENT SENSOR

A. Structure and Operational Principle

Fig. 1 shows the structure of the proposed current sensor for the buck converter, which consists of a sensing MOSFET transistor M_{ps}, switches M_{s1}, M_{s2}, a single-stage amplifier realized by transistors $M_1 - M_4$, a pMOS source follower M_5, a sense resistor R_s, and a local shunt feedback device M_6 and its bias transistor M_7. The proposed current sensor is used to sense the current I_p through power pMOS transistor M_p when Q=0. Since I_p is the same as the inductor current I_L when Q=0, accurately sensing I_p to provide a sensing voltage V_{sense} is used to regulate the output voltage of the current-mode buck converter. As a result, the gate of the sensing transistor M_{ps} is always connected to ground. Switches M_{s1} and M_{s2} are used to ensure that the voltage v_i would not be shorted to ground during Q=1. Moreover, in the proposed current sensor, the aspect ratio of M_{ps} is designed to be 1000 times smaller than that of M_p and then the drain current I_{ps} through M_{ps} is 1/1000 of I_p if and only if the voltage v_x equal to the voltage v_i. In this case, the single-stage amplifier $M_1 - M_4$, the local shunt feedback device M_6, and source follower M_5 form a negative feedback loop to enforce v_x equal to v_i. The output sensing current I_s flows through R_s is the difference between the sensing current I_{ps} and the bias current I_{bias} passing through transistor M_4, and the bias current I_{bias} flowing through R_s proportional to and Small I_{bias} can allow I_s substantially smaller than I_p such that the accuracy of the current sensing can be improved. In the proposed design, small I_{bias} is used to achieve accurate current sensing, while the local shunt feedback enhances the loop response and the speed of the current sensor and it will be explained in the following.

B. Proposed Local Shunt Feedback

Before illustrating the performance improvement by using the local shunt feedback, we need to first understand the performance limitation of the current sensor without the local shunt feedback. Fig. 2 shows the structure of the current sensor without the local shunt feedback, namely as conventional current sensor for convenience. The loop-gain magnitude $|T|$ and unity-gain frequency UGF of the feedback loop in the conventional sensor is given by

$$|T| \approx \frac{g_{m3}g_{ms}r_{o1}R_{ps}}{(1+g_{ms}R_{ps})\sqrt{I_{bias}}} \propto \frac{g_{ms}R_{ps}}{(1+g_{ms}R_{ps})\sqrt{I_{bias}}} \quad (1)$$

$$UGF = \frac{g_{m3}g_{ms}R_{ps}}{C_a(1+g_{ms}R_{ps})} \propto \frac{g_{ms}R_{ps}\sqrt{I_{bias}}}{C_a(1+g_{ms}R_{ps})} \quad (2)$$

where g_{m3} and g_{ms} are transconductances of M_3 and M_{ps}, R_{ps} is the on-resistance of M_{ps}, C_a is the total capacitance at node v_a; and r_{o1} is the output resistance of M_1.

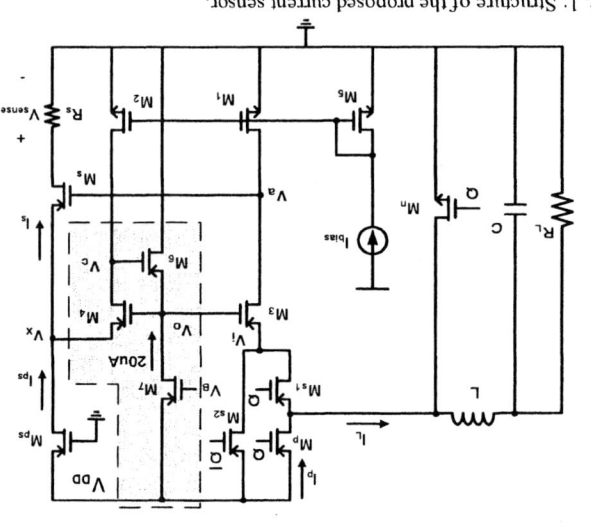

Fig. 1: Structure of the proposed current sensor.

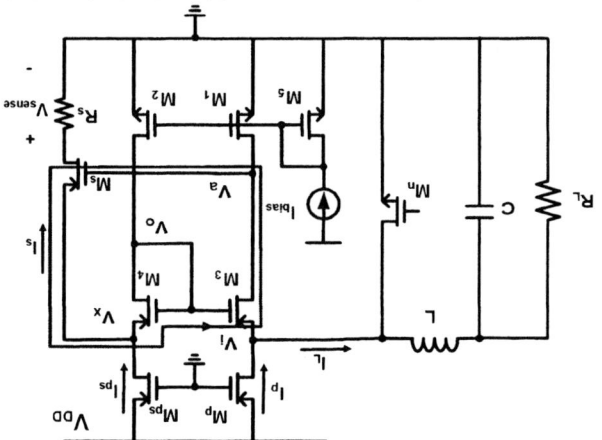

Fig. 2: Structure of the current sensor without local shunt feedback (conventional current sensor) when Q = 0.

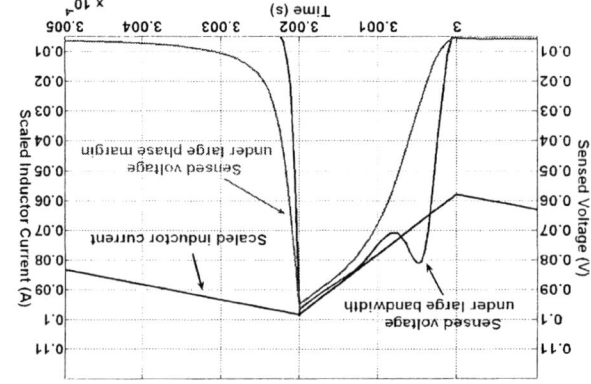

Fig. 3: Accuracy of the conventional current sensor under large bandwidth and large phase margin conditions.

From (1) and (2), both UGF and $|T|$ increase with the value of g_{ms}, which implies that both accuracy and speed of the current sensor can be improved by increasing g_{ms}. However, the speed of the current sensor also depends on the phase margin of the feedback loop that in turn controlled by the location of the non-dominant pole p_{nd}. When p_{nd} is very close to UGF, the phase margin of the feedback loop is decreased such that the settling time of the current sensor is increased to limit the speed of the current sensor. Fig. 3 shows the simulation results of the conventional current sensor under different g_{ms}. When g_{ms} is small, small UGF and |T| limit the speed and accuracy of the current sensor. When g_{ms} is increased, the accuracy of the current sensor is improved. However, the transient overshoot due to the poor phase margin degrades the speed of the current sensor though UGF is large. In order to achieve both high speed and good sensing accuracy, we need to keep using large g_{ms} to maximize UGF and |T| and at the same time improving the phase margin of the feedback loop. In fact, the phase margin (PM) is given as

$$PM = 90° - \tan^{-1}(\frac{UGF}{p_{nd}}) = 90° - \tan^{-1}(R_o \cdot C_o \cdot UGF) \quad (3)$$

where R_o and C_o are equivalent resistance and capacitance at node v_o and $R_o = 1/g_{m4}$ in Fig. 2. From (3), the phase margin can be improved by decreasing the negative phase shift caused by the non-dominant pole.

With the local shunt feedback by using transistor M_6 in Fig. 1, the non-dominant pole of the feedback loop in the proposed current sensor is pushed to a much higher frequency, as the equivalent resistance R_o at node v_o in Fig. 1 is reduced compared to that in Fig. 2. In order to illustrate the reduction of R_o, the small-signal model for calculating R_o is shown in Fig. 4, in which R_o is calculated by injecting a testing voltage v_{ex} into node v_o. The KCL equations for calculating R_o are

$$i_{ex} = \frac{v_{ex}}{r_{o6}} + g_{m6}(v_{ex} - v_c) \quad (4)$$

$$\frac{v_x}{R_{ps}} = \frac{v_c - v_x}{r_{o4}} + g_{m4}(v_{ex} - v_x) \quad (5)$$

$$\frac{v_c}{R_{ps}} = -\frac{v_c}{r_{o2}} \quad (6)$$

By solving (4)-(6) and taking $R_{ps} << r_{o2}$, $g_{m4}R_{ps} >> 1$ and $r_{o6} >> 1$ into considerations, R_o is calculated as

$$R_o = \frac{v_{ex}}{i_{ex}} = \frac{1}{g_{m6} + g_{m6}g_{m4}(r_{o2}//r_{o4})} \quad (7)$$

Obviously, R_o in the proposed current sensor from (7) is at least $g_{m6}(r_{o2}//r_{o4})$ times smaller than that of the conventional current sensor. As a result, the local shunt feedback reduces the negative phase shift caused by p_{nd} and improves the phase margin of the feedback loop according to (3). The proposed current sensor can simultaneously achieve high speed and good sensing accuracy by using the local shunt feedback. Fig. 5 shows the simulated frequency response of negative feedback loops in both conventional and proposed current sensors. As the non-dominant pole has been pushed to a much higher frequency in the proposed current sensor, the phase margin is improved from 50° to 80° and the unity-gain frequency is also increased from 11.5MHz to 14.8MHz. The transient overshoot can thus be eliminated in the proposed current sensor without increasing the biasing current, which is illustrated in the simulated transient responses of both conventional and proposed current sensors in Fig. 6. Moreover, since both

conventional and proposed current sensors can achieve the same loop-gain magnitude in Fig. 5, it proves that the speed of the proposed current sensor is improved without the degradation in the sensing accuracy.

III. SIMULATION RESULTS

A current-mode buck converter with the proposed current sensor has been implemented in a standard 0.35-μm process. The performance summary of the proposed current sensor is given in Table I. In particular, Fig. 7 shows the simulated transient responses of the proposed current sensor under different duty ratios and load currents. From the results, the proposed current sensor can operate at the switching frequency of 2MHz, achieve at least 95% sensing accuracy and only consume 35μA quiescent current under different conditions.

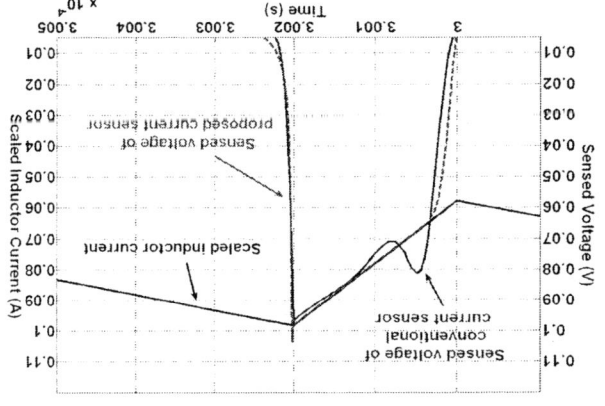

Fig. 4: Small signal model for calculating resistance at node v_o.

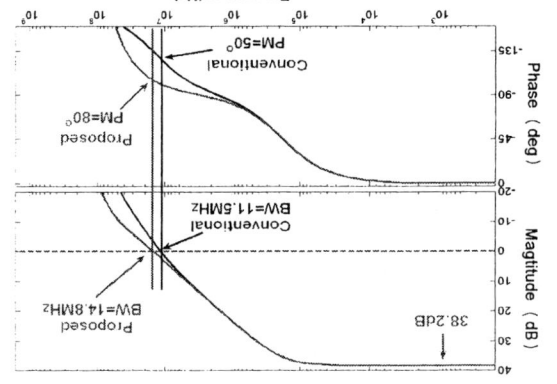

Fig. 5: Simulated loop responses of negative feedback loops in conventional and proposed current sensors.

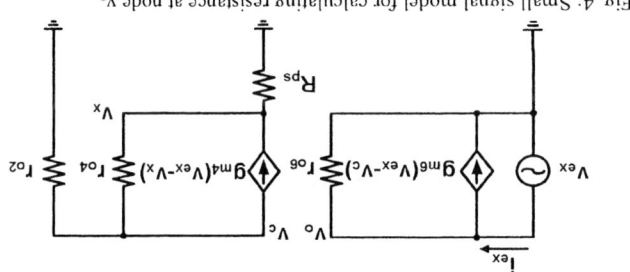

Fig. 6: Simulated transient responses of conventional and proposed current sensors.

Fig. 8 shows the simulated output ripple voltage and inductor current of the current-mode buck converter by using the sensed voltage from the proposed current sensor for the feedback control. In the current-mode buck converter, an off-chip inductor of 2.2μH and an output capacitor of 4.7μF with an equivalent series resistor of 100mΩ are used in the power stage for the power conversion. The buck converter is stable under current-mode operation and small output ripple of 13mV and inductor current ripple of 125mA are obtained. Both results from Figs. 7 and 8 prove the effectiveness of the proposed local shunt feedback to improve the performances of the on-chip current sensor and current-mode buck converter.

IV. CONCLUSIONS

A new on-chip current-sensor with a compact local shunt feedback structure for CMOS current-mode buck converter has been introduced, analyzed and verified in this paper. The local shunt feedback pushes the non-dominant pole in the current sensor to a high frequency such that both speed and accuracy of the current sensor can be improved simultaneously by only consuming low quiescent current. The proposed current sensor is suitable for the current-mode buck converter operating at high switching frequencies in battery-operated portable electronic devices.

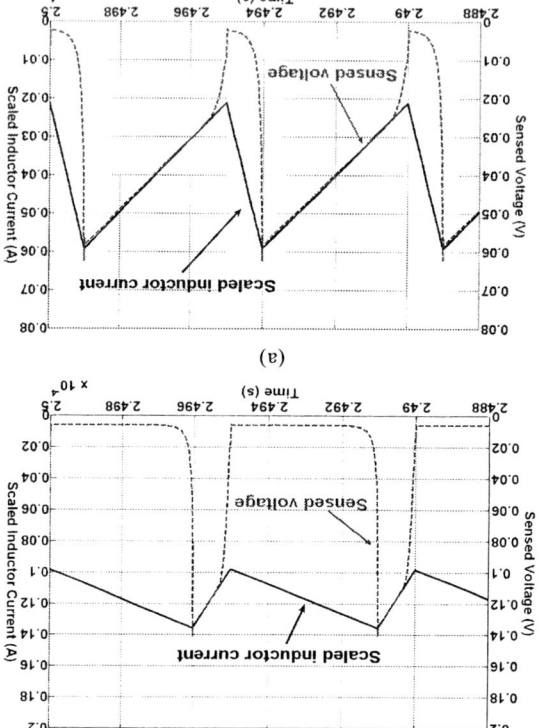

Fig. 7: Sensed voltage and inductor current at switching frequency of 2MHz under (a) duty cycle = 0.2 and load current = 300mA and (b) duty cycle = 0.8 and load current = 100mA.

REFERENCES

[1] W. H. Ki, Current sensing technique using MOS transistors scaling with matched bipolar current sources, U.S. Patent 5,757,174, May 26, 1998.

[2] C. F. Lee and P. K. T. Mok, "On-chip current sensing technique for CMOS monolithic switch-mode power converters," in Proc. IEEE Symp. Circuits and Systems, May 2002, pp. 265 - 268.

[3] C. F. Lee and P. K. T. Mok, "A monolithic current-mode CMOS DC-DC converter with on-chip current-sensing technique," IEEE J. Solid-State Circuits, vol. 39, no. 1, pp. 3 - 14, Jan. 2004.

[4] C. Y. Leung, P. K. T. Mok, and K. N. Leung, "A 1.2-V buck converter with a novel on-chip low-voltage current-sensing scheme," in Proc. IEEE Symp. Circuits and Systems, May 2004, pp. 824 - 827.

[5] C. Y. Leung, P. K. T. Mok, K. N. Leung, and M. Chan, "An integrated CMOS current-sensing circuit for low-voltage current-mode buck regulator," IEEE Trans. Circuits Syst. II, vol. 52, no. 7, pp. 394 - 397, Jul. 2005.

[6] H. Y. H. Lam, W. H. Ki, and D. Ma, "Loop gain analysis and development of high-speed high-accuracy current sensors for switching converters," in Proc. IEEE Symp. Circuits and Systems, May 2004, pp. 828 - 831.

[7] H. P. Forghani-zadeh and G. A. Rincon-Mora, "An accurate, continuous, and lossless self-learning CMOS current-sensing scheme for inductor-based DC-DC converters," IEEE J. Solid-State Circuits, vol. 42, no. 3, pp. 665 - 679, Mar. 2007.

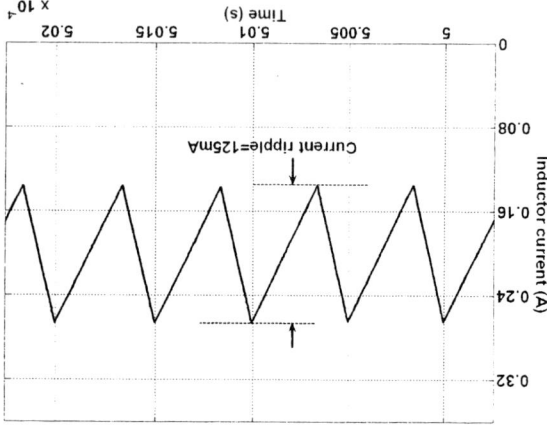

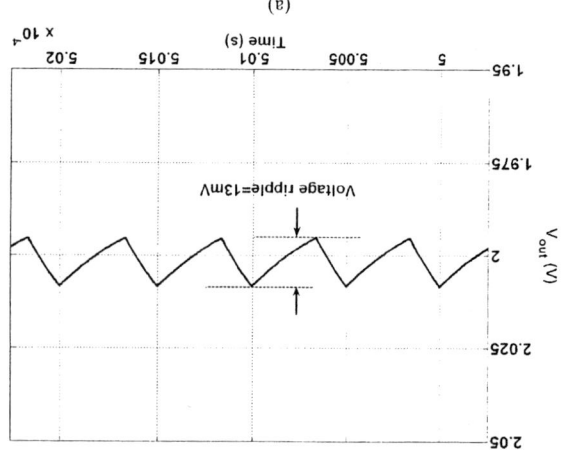

Fig. 8: Simulated (a) output voltage and (b) inductor current of the current mode buck converter with the proposed current sensor.

Author Index

Agarwal R.P – 55
Albina M. Cristian – 43, 71
Atalla Essam – 63

Balsara Poras – 63, 121
Bashir Imran – 63
Buss, Dennis – 1

Chen Wickham – 125
Clynes Steve – 75

Du Mengmeng – 129

Eliezer Oren – 51

Faezipour Miad – 85
Fan Yanli – 125
Feinstein David Y. – 81
Friedman Ofer – 51

Charpurey Ranjit – 47, 101, 105
Ghavami Behnam – 67
Ghosh Diptendu – 101, 105
Gong Datun – 125
Guarisco Michael – 89
Gui Ping – 125
Gulati Kush – 17
Gurrapu Srikanth – 75

Hackl Günther – 43, 71
Haider Towfique – 75
Han Junghwan – 47
Hariyama M. – 59
Hu Jiang – 93
Hu Shiyan – 93
Huang Huanzhang – 125

Joshi R.C – 55

Kameyama M. – 59
Katz Ran – 51
Kaushik B.K – 55
Khan Mohammed Saif – 97
Kiasaleh Kamran – 63
Kimball D. – 39
Kocan Fatih – 81
Kundert Ken – 25

Larson L.E – 39
Lee Hoi – 129
Lie Donald Y.C – 39
Liu Tiankuan – 125

Modi Sankalp S. – 121
Mohammed Wahed – 75
Morgan Mark – 125
Muthumala H.W – 59

Nikolic Bora – 33
Nourani Mehrdad – 85

Parikh Viral – 121
Pedram Hossein – 67
Perrott Michael – 9
Popp J.D – 39

Rabah Hassan – 89

Sarkar S. – 55
Staszewski R.B – 63, 109, 113

Tarim Nil – 117
Thornton Mitchell A. – 81
Tsfati Yossi – 51
Tzoreff Yaniv – 51

Uyanik H.U – 117

Wang. F – 39
Weber Serge – 89

Xiang Annie C. – 125
Xiu Liming – 75
Yanduru Naveen K. – 97
Zhang Xun – 89
Zhu Peiqing – 125

9781424416790